.23

Know Your OSCILLOSCOPE

by

Paul C. Smith

Revised by

Robert G. Middleton

Howard W. Sams & Co., Inc.

4300 WEST 62ND ST. INDIANAPOLIS, INDIANA 46268 USA

THIRD EDITION
THIRD PRINTING—1978

International Standard Book Number: 0-672-21102-5
Library of Congress Catalog Card Number: 74-15451

Printed in the United States of America.

Preface

This book has been prepared for all users of oscilloscopes. The approach is from a technical viewpoint, but the subject matter does not require an engineering background on the part of the reader in order to be understood.

As the title suggests, the reader is first introduced to the principal circuits in an oscilloscope and the function of each. The various accessories available for use with oscilloscopes are then described, along with their special functions. One chapter is devoted to the maintenance and proper adjustment of the oscilloscope, since a defective scope, sitting unused on a shelf and gathering dust, is certainly no asset. The last four chapters in the book describe many of the countless applications of oscilloscopes in the field of electronics.

In this third edition, discussions are brought up to date on the latest scope developments. Transistorized amplifiers and sweep oscillators are explained. Fundamentals of triggered-sweep and dual-trace scopes are discussed. Basic regulated power supplies are presented, with explanation of their operation. Refinements such as edge-lighted graticules and internal graticules are illustrated. Balanced vertical inputs are covered, with an example of application. Calibration of triggered sweeps is discussed. Thus, this new edition provides the technician with a broad spectrum of modern scope design and capabilities. Although the experienced technician may prefer to read individual chapters, the beginner is advised to proceed systematically because early chapters explain basic concepts that are taken for granted in following chapters.

Contents

Chapter 1

General Information

Man depends on his senses to tell him what is going on in the world about him, and he probably depends most on his sense of sight. This accounts, in part, for the popularity and usefulness of the oscilloscope. The oscilloscope provides the service technician with a "third eye," enabling him to see what is happening in the many electronic circuits with which he works.

At one time the oscilloscope was less common, being found mainly in experimental or developmental laboratories. Its use has spread now, and a form of the oscilloscope can be found in practically every radio and tv service shop or almost every industry concerned with electronics.

The appearance of a basic oscilloscope is seen in Fig. 1-1. Although numerous controls are provided on the front panel, the instrument is not difficult to operate. To anticipate subsequent discussion, a half dozen, or even more, of the controls may remain at reference settings during a series of tests. Therefore, the apprentice technician should not jump to the conclusion that an oscilloscope is more difficult to operate than a tv receiver, for example. In fact, a service-type oscilloscope is easier to operate than a color-tv receiver.

The word "oscilloscope" can be separated into two parts. "oscillo" and "scope"; the first is short for "oscillations," and the second means "to view or see." Thus, if we take the word literally, it describes an instrument for viewing oscillations. The term oscillation should be extended to include any vibration or change in amplitude. This applies not only to electrical changes but also to mechanical changes, pressure changes, temperature changes, and so on. Any nonelectrical phenomena must first be converted to electrical signals by means of a transducer, and these signals can then be applied to the oscilloscope.

Some examples of usable transducers are crystal, ceramic, and magnetic pickups and photocells. In most radio and tv applications, an electrical signal is already present or is supplied by accessory equipment. Transducers are not required in such cases—the oscilloscope can be connected directly to the circuits under observation.

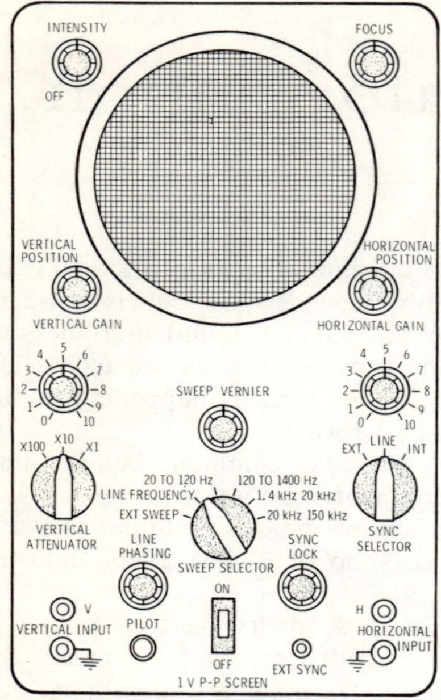

Fig. 1-1. Appearance of the basic oscilloscope.

When the oscilloscope is properly connected and adjusted, it gives the technician a visible indication of the amplitude, frequency, phase, and waveform of the signal at any particular point in a circuit. An instrument providing as much information as this to the knowledgeable user, is a powerful tool indeed. There is probably no phase of electronics where it has not proved useful for designing, testing, or servicing.

EARLY DEVELOPMENTS

The forerunner, or "ancestor," of the oscilloscope was an instrument known as the *oscillograph*. This was a mechanical device for recording oscillations or other natural phenomena of a variable nature. Recordings were made by a pen that left a trace on a moving ribbon of paper, or by the action of a moving light beam upon photosensitive material, or by some other recording means. An example of a mechanical oscillograph is the barograph, which records barometric pressure changes. The recording arm of a barograph or other mechanical oscillograph possesses an appreciable mass, thus limiting the response to frequencies below about 10,000 hertz.* The response can be made quite good, however, down to very low frequencies, and in this one respect at least, the mechanical oscillograph excels the general-purpose oscilloscope.

The mechanical oscillograph would be of little use in radio and television applications, which deal with frequencies in megahertz rather than kilohertz. The modern oscilloscope, however, can be used for such applications. It was made possible by the development of the cathode-ray tube, which will respond to these higher frequencies.

The oscilloscope is really a voltmeter, but it is a voltmeter with special properties. The voltage applied to its terminals determines the position of the electron beam in the cathode-ray tube. The electron beam produces a spot of light wherever it strikes the fluorescent surface of the tube. As the beam moves across the face of the tube in response to the voltages on the deflection plates, the spot of light also moves; if the movement is fast enough, the spot appears as a continuous trace or line of light.

PERSISTENCE

This blending of successive positions of the spot into an apparently continuous trace is due to two factors: (1) the persistence of the phosphor of the tube and (2) the persistence of vision (the property of the human eye that sees any object or spot of light at its original position for a fraction of a second after it has moved). The persistence of a phosphor is its brief glow after the electron beam has left that spot. In general-purpose oscilloscopes, the blending of the spot into a line is due almost entirely to the persistence of vision. In special-purpose oscilloscopes, a tube with a phosphor of long persistence may be used, and electrical phenomena of short duration and nonrepeating nature can be viewed.

The phosphor most commonly used in oscilloscopes is P1, rated at medium persistence. P5 is a phosphor of short persistence, and P7, one of long persistence. Other phosphors are P4, commonly found in tv picture tubes, and P11, which gives an easily photographed blue trace.

---

* One hertz equals 1 cycle per second. Thus 10,000 hertz (10 kilohertz) equals 10,000 cycles per second.

GRAPH PATTERNS

To use a simple analogy, the electron beam can be considered as a pencil writing upon the screen of the cathode-ray according to the voltage on the deflection plates. When a horizontal-deflection system is used (and practically no oscilloscope is built without one), the trace on the screen is really a graph. Graphs are now so commonplace that hardly a person has not seen one. Some examples are the temperature graphs and electrocardiographs used in hospitals, or the sales graphs of a business office.

The reader probably has drawn graphs in school; he will remember that they show two sets of data, the values of one set varying in some fashion as the other set varies. One set of values is plotted along the horizontal or X-axis on the graph paper, and the other set is plotted along the vertical or Y-axis. The located points are then connected to form a continuous graph. The action of the oscilloscope in tracing a response curve is so similar to this that some oscilloscopes even have inputs marked "X-amplifier" and "Y-amplifier."

We believe this comparison is a good point to remember. When confusing indications are seen on the oscilloscope screen, it may help if the operator remembers that the oscilloscope is plotting time horizontally and voltage vertically to produce a graphical account of the operating conditions of the circuit. Usually it is unnecessary to know exactly how much time is represented by the horizontal travel of the trace, as long as the beam is uniform in its rate of travel; but, if necessary, this time can be determined accurately and for very short intervals.

Fig. 1-2 shows a graph of one cycle of voltage having a frequency of 60 hertz. Instantaneous voltage is plotted above or below the X or horizontal axis; elapsed time is plotted to the right of the vertical axis (Y) and measured in fractions of a second. The peak voltage is taken as 1 to simplify plotting the graph. This curve is called a sine curve because the amplitude or Y value at any point on the curve equals the maximum value of E (in this case, 1) times the sine of the X value at that point. (X must be converted to degrees, with one complete cycle equalling 360 degrees.)

Practically all oscilloscopes have a *graticule* mounted in front of the face of the cathode-ray tube. The graticule is commonly ruled like a sheet of graph paper. The vertical intervals are used to measure voltage, and the horizontal intervals are used to measure time in typical applications. Since the plastic graticule is mounted in front of the fluorescent screen in most oscilloscopes, you must view the pattern in a line exactly perpendicular to the screen to minimize parallax error. If you view the screen obliquely, the parallax error can be appreciable in using the graticule. Just as some meters have mirrored scales to mimimize parallax error, so do some modern crt's have parallax-free graticules. For example, the oscilloscope illustrated in Fig. 1-3 has the graticule ruled on the *inner* surface of the crt face.

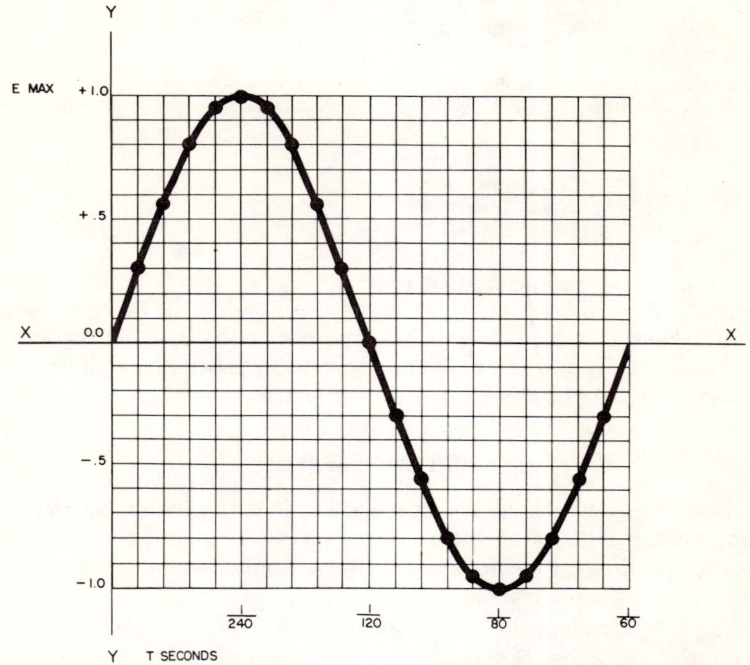

Fig. 1-2. Graph of one cycle of power-line voltage. Frequency is 60 Hz.

Lab-type scopes often have *edge-lighted* graticules. An example of this design is illustrated in Fig. 1-4. The graticule is fabricated from lucite, and several red-light bulbs are mounted around the edge of the graticule. When the scale illumination control is turned down, the graticule is almost invisible, and only the green fluorescent pattern is in evidence. However, when the scale-illumination control is turned up, the graticule rulings are displayed as red lines over the green waveform. The variable graticule illumination facilitates photography of waveforms.

It is interesting to note that the distinction between lab-type and service-type oscilloscopes has become less definite than in the past. As an illustration, Fig. 1-5 shows an elaborate service-type oscilloscope. This instrument has display and operating features which

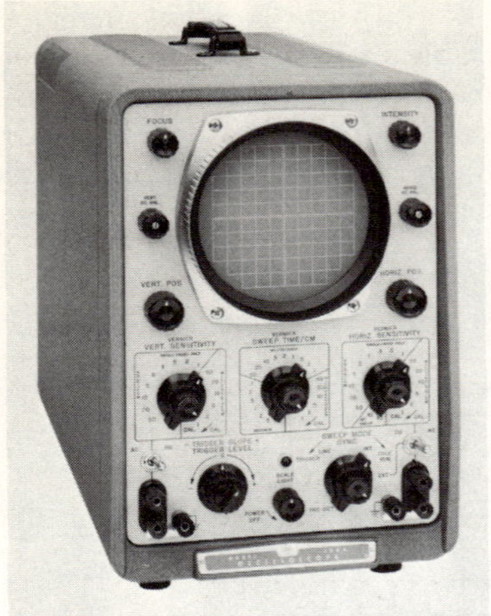

Courtesy Hewlett-Packard

Fig. 1-3. The graticule is ruled on the inside surface of the cathode-ray tube in this oscilloscope.

would have classified it as a lab-type oscilloscope in the not-too-distant past. By the same token, present-day lab-type oscilloscopes are far more sophisticated than in the past.

WRITING SPEED

Before discussing the oscilloscope section by section, an important characteristic of all oscilloscopes should be mentioned—the reaction speed of the electron beam to any applied voltage. The beam

Courtesy Tektronix, Inc.

Fig. 1-4. An oscilloscope with an edge-lighted graticule.

possesses very little inertia. For all practical purposes, it can be said to have no inertia; consequently, it responds almost instantaneously to the impulse of the deflection voltages. This is the property that enables the trace to follow every variation of the applied signal, no matter how suddenly the signal may change direction or amplitude.

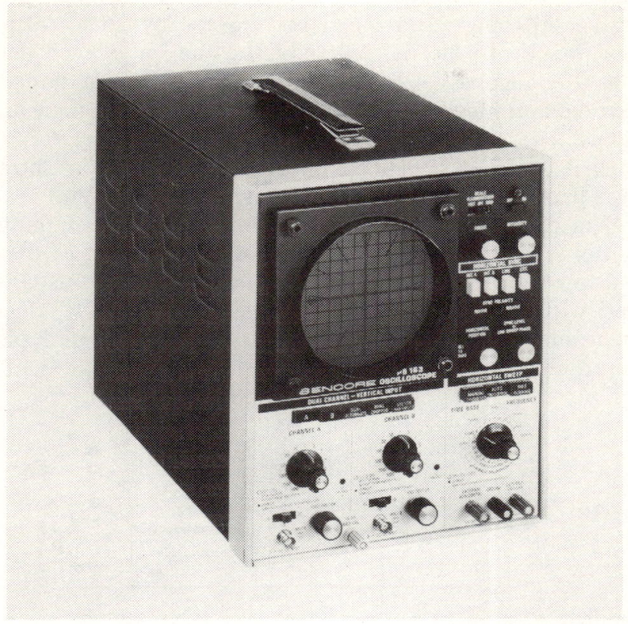

Courtesy Sencore

Fig. 1-5. An elaborate service-type oscilloscope.

How readily the beam changes direction while moving at high speed can be shown by the following example. Assume an oscilloscope has a sweep frequency of 30 kHz and a horizontal amplification capable of expanding the trace to four times the screen width. (Many oscilloscopes will exceed both specifications.) For a 5-inch oscilloscope, this means the trace is equal to 20 inches in length although only 5 inches of the center can be seen. The beam sweeps these 20 inches in 1/30,000 second—actually, in even less time since some time is lost in retrace. Thus, the beam is sweeping the tube at a "writing speed" of 600,000 inches per second, a little faster than 34,000 miles per hour. The retrace time usually is less than trace time; accordingly, the retrace speed would be much greater. However, the retrace is seldom used for viewing and, therefore, is not considered when discussing writing speed.

INPUT IMPEDANCE

Another important characteristic of the oscilloscope is its high input impedance. This is desirable in any voltage-measuring instrument, for it means the instrument will have a minimum loading or disturbing effect on any circuit to which it is connected. The vertical amplifier input impedance of a conventional oscilloscope may have any value from 1 to 5 megohms shunted by 25 to 50 pF. If connected directly to the deflection plates, the impedance may be as high as 10 megohms shunted by 15 pF. The input impedance at the vertical amplifier can be increased by the use of high-impedance probes.

A block diagram of a general-purpose oscilloscope is shown in Fig. 1-6. This is a greatly simplified diagram with several features combined in each block. The focus, intensity, and positioning circuits are not shown; they have been considered as part of the low-voltage power supply. The step and vernier attenuators usually are associated with the vertical and horizontal amplifiers. Triggering and synchronizing of the sweep oscillator are considered part of the sweep oscillator.

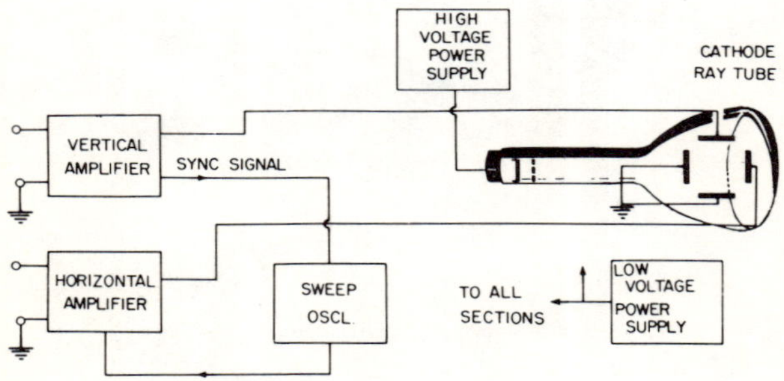

Fig. 1-6. Simplified block diagram of a general-purpose oscilloscope.

An oscilloscope could be made of a cathode-ray tube and a power supply only. Such an oscilloscope would be extremely limited in the ways it could be used. The signal input would have to be made directly to the deflection plates, and a comparatively strong signal would be necessary to deflect the electron beam a usable amount. After adding vertical and horizontal amplifiers and a horizontal-deflection system to provide a time base, the oscilloscope may be used for an increased number of applications. The oscilloscope can respond to very weak input signals, and general-purpose oscilloscopes sometimes have a vertical-deflection sensitivity of 15 millivolts rms per inch or less.

THE CATHODE-RAY TUBE

A modern 5-inch cathode-ray tube is shown in Fig. 1-7. Externally, it has four parts: the base, the neck, the bulb, and the face or screen. Inside the neck, a portion of the gun structure can be seen. Fig. 1-8 shows the gun structure removed from the tube. The gun contains all the electrodes for forming, shaping, and directing the electron beam which strikes the fluorescent screen of the tube.

Fig. 1-7. Modern 5-inch cathode-ray tube.

Applying the proper voltage to the various electrodes of the gun produces a beam that is brought to a focus in a small spot on the tube screen. The beam intensity is controlled by the voltage on the control grid. The theory pertaining to the focusing action of the gun is probably less interesting to the service technician than the

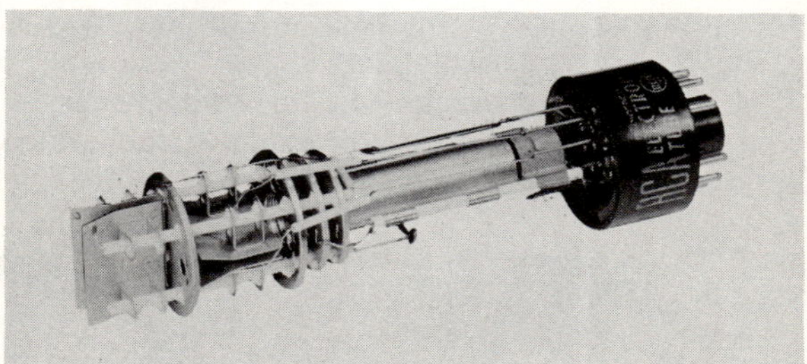

Fig. 1-8. Electron gun of cathode-ray tube.

theory pertaining to the action of the deflection plates; consequently, more space will be devoted to the latter. This book will not deal with electromagnetic deflection systems since they are found almost exclusively in television receivers rather than in oscilloscopes. Fig. 1-9 is a perspective drawing showing how the electron beam passes through the space between the deflection plates on its path to the screen. With all deflection plates at the same

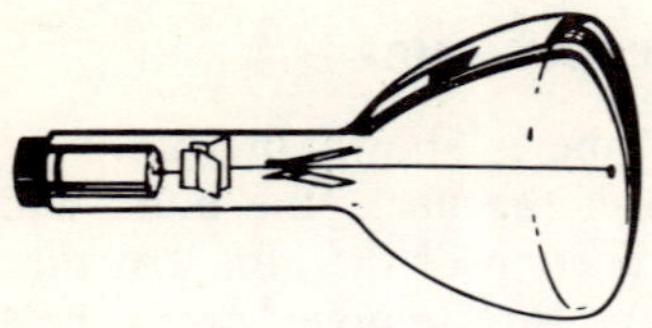

Fig. 1-9. Path of electron beam through deflection-plate assembly.

electrical potential, the beam will pass along the axis of the deflection-plate assembly and strike the center of the screen.

If one plate of a pair of deflection plates is made more positive or negative than the other, the electron beam is attracted toward the positive plate and repelled from the negative plate (Fig. 1-10), because unlike electron charges attract and like charges repel each other. The electron beam is always negative and therefore is always attracted to the positive plate. The amount of deflection varies directly with the magnitude of the voltage on the deflection plates. For example, if a potential difference of 50 volts between a pair of plates moves the beam one inch at the screen, 100 volts will move it two inches, and so on. This is shown in Fig. 1-11.

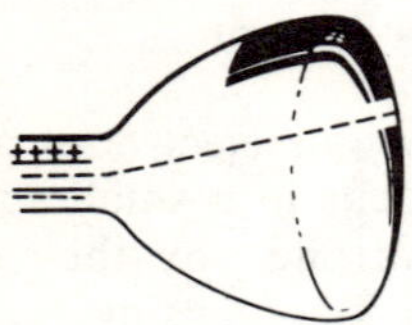

Fig. 1-10. The electron beam, being negative, is always attracted by the positively charged deflection plate and repelled by the negatively charged deflection plate.

Applying an alternating voltage to the vertical plates moves the beam and produces a vertical line from top to bottom of the screen. Similarly, the proper voltage applied to the horizontal plates produces a horizontal line across the screen. With proper voltages for both sets of plates, the beam can be made to move anywhere on the oscilloscope screen.

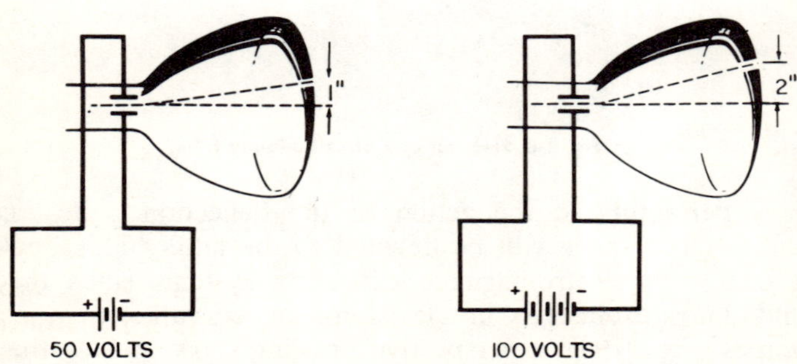

Fig. 1-11. The amount of deflection is directly proportional to the voltage applied to the plates.

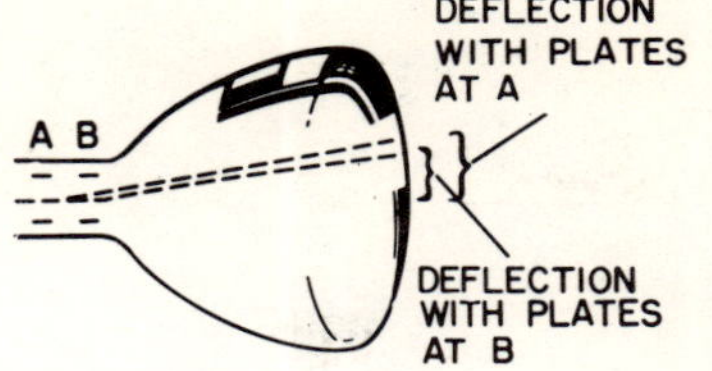

Fig. 1-12. Given equal deflection voltages, deflection plates nearer the base of the cathode-ray tube will have greater deflection sensitivity than those farther away.

DEFLECTION SENSITIVITY

Deflection sensitivity of a cathode-ray tube, and of the entire oscilloscope, determines the weakest signal than can be viewed successfully with the instrument. Anyone who has consulted a tube manual about cathode-ray tubes may have noticed that deflection sensitivities can cover a wide range, depending on the voltages used. The sensitivities also differ for the two pairs of deflection plates, one sensitivity being greater than the other. For example, one tube manual lists the following sensitivities for a 5CP1-A cathode-ray tube. When the voltage of anode No. 3 is twice that of anode No. 2, the sensitivity is 39 to 53 volts (dc) per inch for every thousand volts supplied to anode No. 2. This range applies to one set of deflection plates. For the other set under the same voltage conditions, the sensitivity is 33 to 45 volts (dc) per inch per thousand volts supplied to anode No. 2. A different set of sensitivity figures is listed for the tube when anodes No. 2 and No. 3 have equal voltages.

The pair of deflection plates having the greater sensitivity (that is, requiring the smaller number of volts per inch of deflection) is always the pair nearer the base of the tube. The reason can be readily seen by examining Fig. 1-12. In this illustration a pair of deflection plates is shown in two different positions, one (position A) being nearer the tube base than the other. If the applied voltage is the same for both positions, the electron beam will be deflected through an equal angle each time. With equal deflection angles, the deflection plates at position A swing a longer beam, thus giving a longer trace on the screen for the same deflection voltage.

Either pair of plates can be used for the vertical system; the rotational position of the tube about its long axis determines which pair. To obtain the highest possible deflection sensitivity for the vertical system, the cathode-ray tube normally is so positioned that the pair of plates closer to the base produce vertical deflection. The horizontal-deflection plates usually are driven by a stronger signal and are farther from the base than the vertical-deflection plates.

Chapter 2

Power Supplies

The power requirements of a modern oscilloscope are usually met with a power supply having two sections—one of low dc voltage (about 300 volts) and medium current capabilities, the other of comparatively high dc voltage (1000 volts or higher) and low current capabilities. The low-voltage section operates the amplifiers and deflection generator. The high-voltage section furnishes the potentials for the various elements of the cathode-ray tube. The power supply also may deliver signals for certain types of synchronization, retrace blanking, and calibration. To gain a better picture of some of the demands upon the power supply, let us first discuss certain aspects of the cathode-ray tube.

ELECTRON PATH THROUGH A CATHODE-RAY TUBE

Fig. 2-1 shows a 5UP1 cathode-ray tube as it commonly appears in the schematic diagrams of oscilloscope instruction books. The order indicated for the elements, starting from the base of the tube, is the same in the diagram as for the actual tube, except possibly anode No. 2. Anode No. 2 is connected internally to grid No. 2. It is also connected to the coated interior of the bulb of the tube, although this is not shown in the diagram. The coating extends almost to the face of the screen and accelerates the electron beam on its way to the screen, also collecting the electrons of the beam after they have struck the screen.

The path of the electrons through the cathode-ray tube is as follows. The electrons are emitted from the heated cathode and, being negative, are attracted toward the nearest positive element, grid No. 2. They pass on through apertures in grid No. 2 and anodes No. 1 and No. 2 and are subjected to the action of the deflection plates. Finally, they strike the screen of the tube, causing a spot or trace of light, and they then are collected by the interior

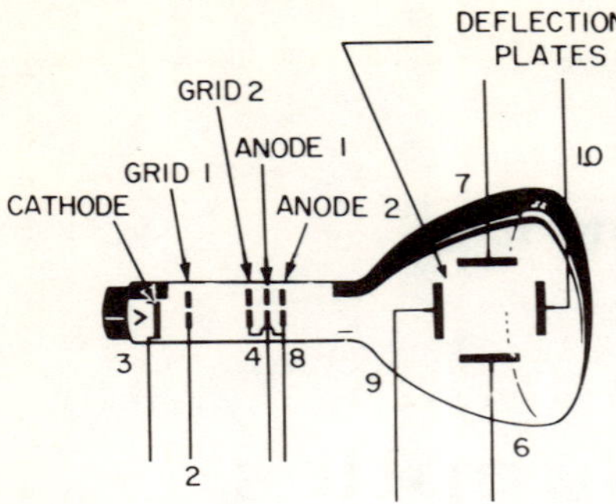

Fig. 2-1. Control elements in a 5UP1 cathode-ray tube.

coating which forms a part of anode No. 2. Thus, the electron path through the tube originates at the cathode and terminates at anode No. 2.

Polarities of the Tube Elements

Proper operation of the cathode-ray tube requires anode No. 1 to be more positive than the cathode and anode No. 2 to be more positive than anode No. 1. Grid No. 1 is the control grid and operates at a voltage equal to or more negative than that of the cathode. Its action is similar to that of the control grid of a receiving tube—it controls the number of electrons flowing between cathode and anode. Being negative, it repels the negative electrons, and if it becomes negative enough, the electron beam is cut off entirely.

The intensity control of the oscilloscope is usually connected to grid No. 1 of the cathode-ray tube, although it may be connected to the cathode instead. A variable intensity depends on a variable potential difference between the cathode and grid, and this variable potential difference can be obtained if the potential of one element is varied while the potential of the other is held constant. The technician is familiar with this aspect of the operation of the cathode-ray tube through his association with television receivers. In some receivers, the picture-tube element to which the brightness control is connected is the control grid; in others, it is the cathode.

Range of Voltage for Normal Operation

As was stated previously, the necessary potentials for operation of the cathode-ray tube are furnished by the high-voltage section of the power supply. These potentials can vary over a wide range, and satisfactory operation will still be obtained. For example, the voltage at anode No. 2 of a 5UP1 tube can be from 1000 to 2500 volts with respect to the cathode. Operation below 1000 volts is not recommended. No matter which voltage is chosen, there are some advantages and disadvantages. The lower voltages can be

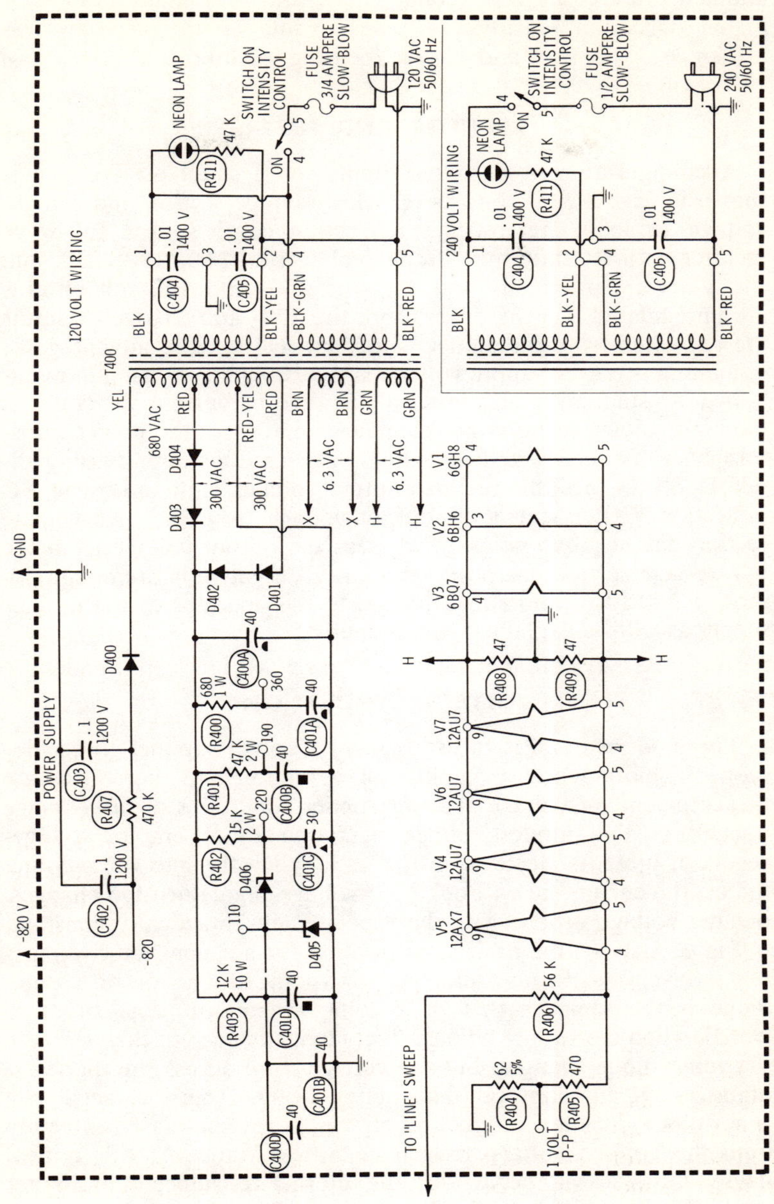

Courtesy Heath Co.

Fig. 2-2. Configuration of a power supply for a small oscilloscope.

attained more easily and economically, and they make possible a higher deflection sensitivity. These advantages are offset by less brilliance of the spot and poorer focusing qualities.

RECTIFIER AND FILTERS

A configuration of a power supply for a small oscilloscope is shown in Fig. 2-2. It consists of a low-voltage section and a high-voltage section. The low-voltage section consists of a full-wave rectifier with RC filtering. Output voltages of 360, 190, 220, and 110 volts are provided. Note that the 360- and 190-volt outputs are unregulated. On the other hand, the 220- and 110-volt outputs are regulated by zener diodes D405 and D406. The advantage of regulating a power-supply voltage is that its voltage value is maintained constant over an appreciable range of current demand. In the high-voltage section, the secondary winding of the power transformer is also connected to a half-wave rectifier circuit using diode D400. In turn, the rectified output voltage is filtered by an RC pi section. Observe that the output voltage from the high-voltage section has negative polarity, whereas the output voltage from the low-voltage section has positive polarity. Negative high-voltage polarity is used to avoid shock hazard to the operator of the oscilloscope, as will be explained subsequently.

VOLTAGE REGULATION

The modern trend is to use regulated power supplies in oscilloscopes. This makes for stability of operation and more accurate measurements of voltage and time under conditions of line-voltage fluctuation. The simplest voltage-regulating circuits employ voltage-regulator tubes, as depicted in Fig. 2-3. When the line voltage fluctuates, the current flow through the voltage-regulator tube changes, but the voltage drop across the tube remains practically constant.

Fig. 2-4 shows the basic configuration for a zener diode voltage-regulator circuit. From a practical viewpoint, the action of a zener diode is the same as that of a voltage-regulator tube. In other words, when the supply voltage fluctuates, the current flow through the zener diode changes, but the voltage drop across the diode remains practically constant. This action results from the fact that a zener diode does not conduct current in its reverse-biased direction until the applied voltage rises to a critical value. At this critical value, "breakdown" occurs and the internal resistance of the zener diode becomes very small. Note resistor R in Fig. 2-4; this is a current-limiting resistor which prevents the zener diode from drawing so much current from the source that the diode would be damaged. Note also that the load resistance, R_L, may vary over an appreciable range, and the voltage across the zener diode will remain practically unchanged.

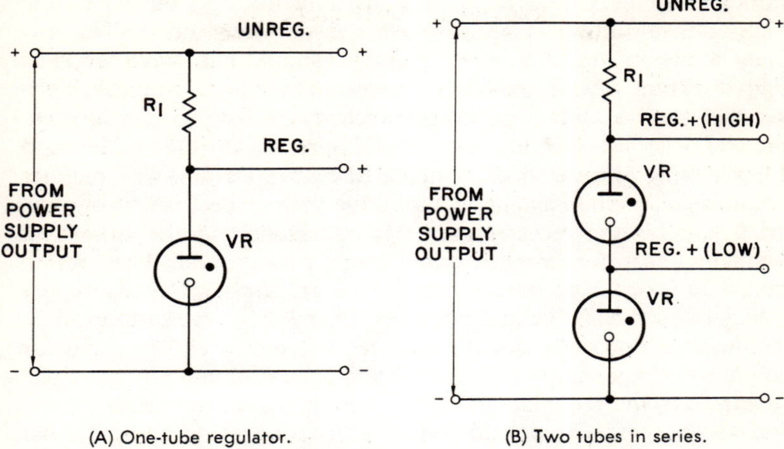

(A) One-tube regulator. (B) Two tubes in series.

Fig. 2-3. Basic configuration for voltage-regulator tubes.

You will find transistorized regulated power supplies in many lab-type scopes. The principle of operation is basically the same as in tube-type regulated power supplies. In other words, the voltage reference is provided by a voltage regulator tube or zener diode. If the current demand on the power supply increases, the transistors "sense" that the output voltage is decreasing with respect

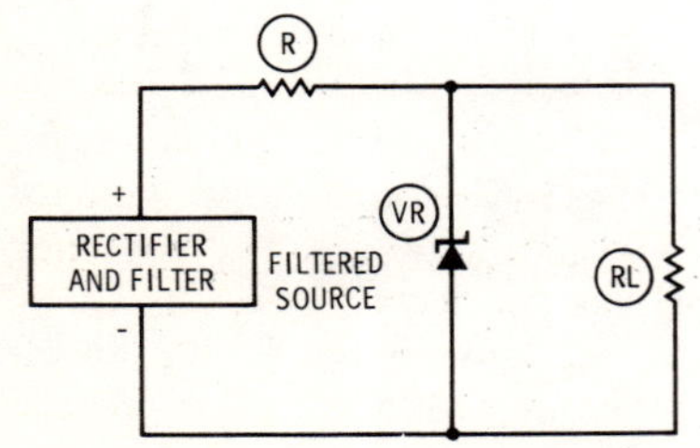

Fig. 2-4. Basic zener diode voltage-regulator circuit.

to the reference voltage. In turn, the base-emitter bias automatically changes to permit more current flow and thereby maintain the output voltage at a practically constant value. The design details of regulated power supplies are extensive, and interested readers are referred to specialized books on the subject.

NEGATIVE HIGH-VOLTAGE SUPPLY

A noticeable feature of the power supply in Fig. 2-2 may seem strange to the person accustomed to the usual power supplies in radios, amplifiers, and many test instruments. The output of the high-voltage section is negative with respect to ground. The circuits shown in Fig. 2-5 are diagrams of simple half-wave rectifiers and illustrate several possible arrangements. Fig. 2-5A shows a rectifier system with a dc output voltage positive with respect to ground. E_s represents the ac voltage impressed on the system, and the resulting electron flow is indicated by the arrows. The rectifier conducts only when its anode is positive with respect to its cathode. In Fig. 2-5B the rectifier has been reversed, and the dc output voltage is therefore negative with respect to ground. The ground connection could be made as in Fig. 2-5C, thus giving dc supply points of both negative and positive polarity with respect to ground.

Most oscilloscopes use the arrangement shown in Fig. 2-5B for the high-voltage supply, often with minor variations. For example, in Fig. 2-6, the rectifiers and filter sections have been omitted; the final stages and the positioning controls for one of the amplifier channels are shown. R1 and R2 form the ground return for one pair of deflection plates, and the ac output from Q1 and Q2 is de-

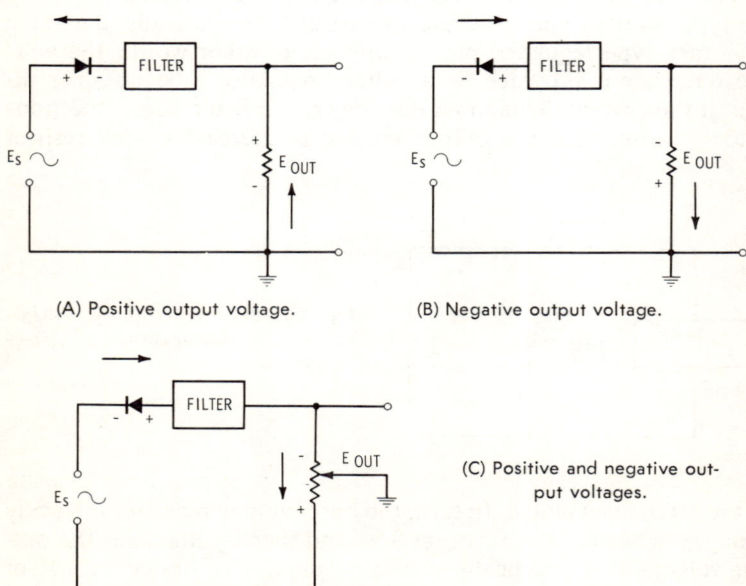

(A) Positive output voltage.

(B) Negative output voltage.

(C) Positive and negative output voltages.

Fig. 2-5. Several variations of half-wave rectifier circuits.

veloped across these two resistors. This is a push-pull deflection system in which a negative-going signal is applied to one deflection plate of a pair at the same time that a positive-going signal is applied to the other plate.

With the circuit arrangement shown in Fig. 2-6, the dc potential of either deflection plate will not vary greatly from ground potential. Any variation will be due to the action of positioning controls R3A and R3B. These controls are ganged and so wired that any rotation of the common shaft shifts the slider of one control toward a more positive potential and, at the same time, shifts the slider of the other control toward a more negative potential. As a result, the deflection plates have push-pull action for the dc positioning voltage as well as for the ac signal.

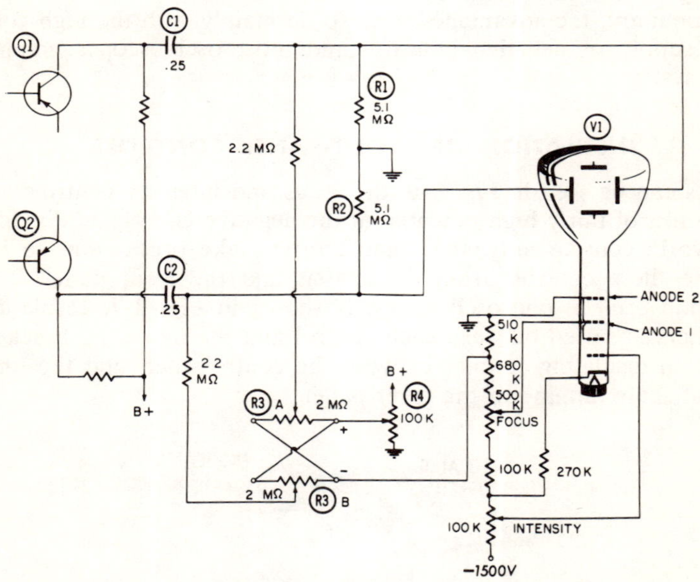

Fig. 2-6. Partial schematic showing the positioning control and high-voltage divider network of an oscilloscope.

The following advantages result from a negative high-voltage supply: (1) The deflection plates can be operated at a dc potential close to that of anode No. 2, thus eliminating the defocusing effect obtained when the two potentials differ greatly. (2) Capacitors C1 and C2 can have a fairly low voltage rating. (3) The circuit can be more easily adapted to dc connection between deflection plates and amplifiers. (4) Less insulation is needed between the positioning controls and the chassis or the front panel.

Contrast the preceding conditions with those obtained if the polarity of the high-voltage supply were reversed: (1) Anode No. 2 would be at a high positive potential to ground, resulting in an extreme difference in potential between the deflection plates and anode No. 2 if dc connections are made from the amplifier to the deflection plates. (This condition is undesirable.) (2) If blocking capacitors C1 and C2 are used, the deflection plates and anode No. 2 would be at nearly the same potential, but the voltage rating of the capacitors would have to be high. Capacitors of that value and rating would be bulky and expensive. (3) The horizontal- and vertical-positioning controls would have to be highly insulated from the chassis and the front panel to protect against the high voltage.

Regardless of the polarity of the high-voltage supply, the voltage rating of filter capacitors C402 and C403 in Fig. 2-2 must be high. In summary, the advantages seem to lie mainly with the high-voltage supply of negative polarity, and most oscilloscopes employ this system.

INSULATION OF FRONT-PANEL CONTROLS

As can be seen in Fig. 2-6, the focus and intensity controls are at points of fairly high potential at the negative end of the dividing network; consequently, the manufacturers take precautions to insulate these controls from the chassis and the front panel. The method used in one oscilloscope is shown in Fig. 2-7. Insulating washers are used between each control and the mounting bracket, with an insulating coupler between the control shaft and the long metal shaft running to the front panel.

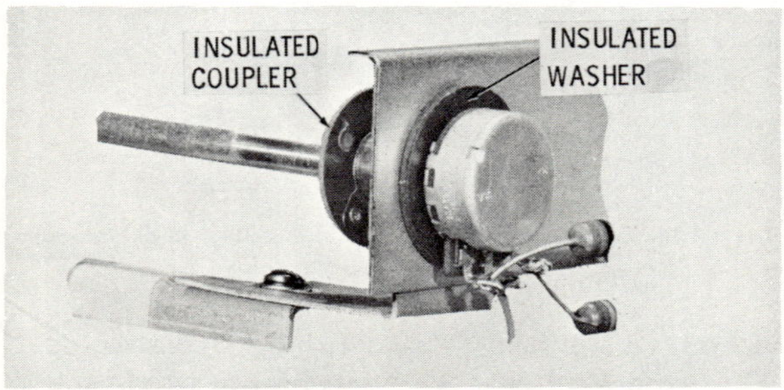

Fig. 2-7. One method of insulating between a control and the chassis.

BEAM INTENSIFICATION

Because it is popular with oscilloscope manufacturers, the 5UP1 cathode-ray tube has been used as an example. However, other cathode-ray tubes are also found in oscilloscopes, and some require a power supply slightly different from those discussed so far. An intensifier anode (called anode No. 3) in the 5ABP1 tube and 5CP1A tube may sometimes be operated at a potential as much as 2000 volts positive with respect to ground; and at the same time, the control grid may be as much as 2000 volts negative with respect to ground. The intensifier anode greatly accelerates the electrons in the beam after they have passed between the deflection plates. A brighter spot results, yet the deflection sensitivity is not seriously affected. The increased velocity of the electrons in the beam permits a higher scanning rate. In order to obtain the positive high voltage for the intensifier anode, another half-wave rectifier system can be added.

With all these high-voltage sources present within the case, the operator should be extremely careful when examining the interior of oscilloscopes. The instruction manuals caution against operating the oscilloscope with the chassis outside its case. Before touching any part of the interior of an oscilloscope, the operator should make sure the filter capacitors are not charged.

Chapter 3

Sweep Systems

In Chapter 1 it was mentioned that the oscilloscope will actually plot a graph of voltage with respect to time. The operator of an oscilloscope can see on its screen an indication of the way a voltage changes in amplitude from one moment to the next. The signal to be observed is normally applied to the vertical deflection system and will cause a vertical trace to appear on the screen, provided the signal is of sufficient amplitude and there is no ac voltage applied to the horizontal-deflection plates.

Under these conditions, a change of amplitude of the signal will result in a change of the heighth of the vertical trace. In order that these changes in amplitude may be viewed with respect to changes in time, some type of sweep system is incorporated in the oscilloscope. The signal from the sweep system is used to drive the horizontal-deflection plates of the cathode-ray tube. This provides a horizontal trace as a time reference for the signal at the vertical-deflection plates. Because of this, sweep systems are sometimes called time bases. In addition to the sweep signals provided internally in the general-purpose oscilloscope, other sweep signals can usually be applied from an external source.

Oscilloscope sweeps may be classed as linear or nonlinear, and as single or repetitive. Single sweeps are seldom found except in laboratory oscilloscopes. Their greatest usefulness is for viewing signals of a nonrecurring nature. They are designed to sweep the beam once across the screen of the oscilloscope and must be timed accurately so that the signal to be viewed will occur at the exact instant of the sweep. A sweep of such short duration would result in a trace that would fade very quickly on a screen of normal persistence; consequently, a screen of long persistence is used to increase the viewing time.

The majority of the signals the service technician will encounter are of a recurring nature. They normally go through a complete cycle of variations a number of times a second. Some examples of this type of signal are: (1) the voltage supplied by the power line, (2) the ac voltages at tube filaments in a receiver, and (3) the voltages generated by the sweep circuits in a television receiver. The ideal sweep for viewing these signals is one in which the beam starts at the left-hand edge of the oscilloscope screen and moves at a uniform rate of speed in a horizontal direction to the right edge of the screen. When it reaches the right edge, the sweep should reverse direction and return to the starting position at the left of the scren. This return sweep (called retrace) should be made in the least time possible.

LINEAR SAWTOOTH SWEEP

The waveform of the voltage necessary to produce such a sweep is shown in Fig. 3-1. Several cycles of the sawtooth waveform are shown here. The voltage applied to the horizontal deflection plates is plotted in a vertical direction, and time is plotted in a horizontal direction. The sweep produced by such a waveform is called a linear sweep because the useful portion of it moves at a constant rate of speed and can be represented by a straight line on a graph. In many oscilloscopes the retrace is blanked out and does not appear on the screen.

Retrace blanking can help prevent some of the confusing indications that might be seen without blanking. This is especially true where some of the higher sweep frequencies are used. Usually, raising the sweep rate higher and higher will result in sweep voltage cycles containing a larger percentage of retrace time. If the retrace is permitted to appear on the screen together with any vertical deflection caused by a signal at that time, it may obscure some more important detail occurring during the forward portion of the sweep.

Blanking can be accomplished by applying the retrace signal from the sweep generator circuits to either the cathode or grid of the cathode-ray tube for intensity modulation. Circuits may be inserted between the generator and cathode-ray tube to provide any waveshaping, phase shifting, or amplification that may be neces-

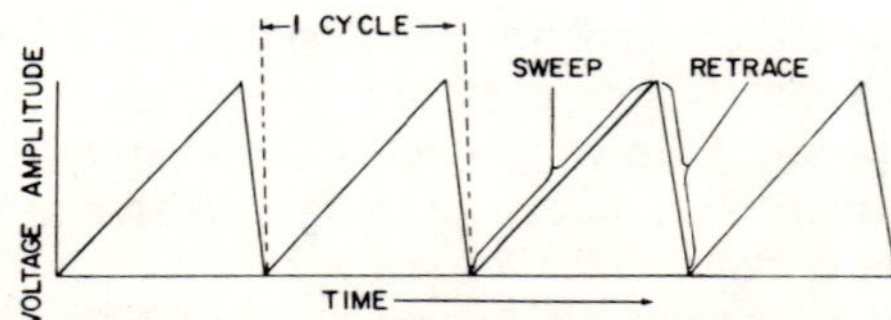

Fig. 3-1. Voltage waveform used to produce a waveform on an oscilloscope screen.

sary. For some applications retrace blanking may not be desirable, and some oscilloscopes are provided with a switch so that the blanking feature may be turned on or off as desired.

There are three common circuits for producing the sawtooth voltage indicated in Fig. 3-1. One of these, the blocking oscillator, is used more in television receivers than in oscilloscopes and will not be discussed here. The other two circuits require the use of a multivibrator or a four-layer diode oscillator. Let us first consider the circuit using the four-layer diode oscillator.

FOUR-LAYER DIODE AS A SWEEP OSCILLATOR

The waveform of Fig. 3-1 can be approximated very closely by the voltage across a capacitor being charged and discharged in a certain manner. Fig. 3-2 shows a simple arrangement for doing this. When the switch is in position A, capacitor C will be shorted, and no voltage will appear across its terminals. When the switch is moved from point A to point B, the battery will immediately start to charge the capacitor and will continue to charge capacitor C until the voltage across the capacitor equals that across the battery. Theoretically it would take an infinite length of time for E_C to reach the voltage E_B. For most practical purposes, E_C can be considered to equal E_B after a time equal to 5RC has elapsed. R-C time is measured in seconds and is equal to the product of the resistance in megohms times the capacitance in microfarads.

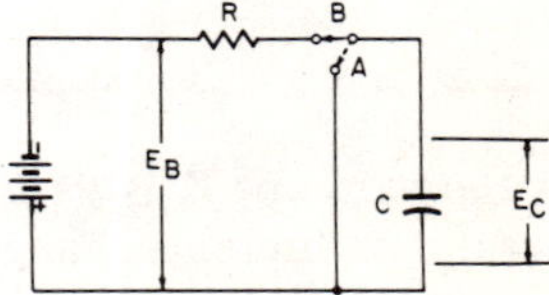

Fig. 3-2. A simple arrangement for charging and discharging a capacitor.

Fig. 3-3 is a graph showing the ratio between the voltage E_C and the voltage E_B obtained with the circuit of Fig. 3-2. It can be seen that the voltage E_C increases rapidly at first, then more slowly as E_C approaches E_B. Considered as a whole, the curve of Fig. 3-3 appears to have a large amount of curvature, but if only a small portion of the curve is considered at one time, it appears to be nearly straight, especially between points 0 and 1RC. It would therefore be logical to use this latter portion of the curve, or a part of it, to develop the sawtooth curve diagrammed in Fig. 3-1. The manner in which this is done can be explained through the use of Fig. 3-4, which illustrates a simple sawtooth oscillator using a four-layer diode. Note that the dc voltage source is applied across an RC

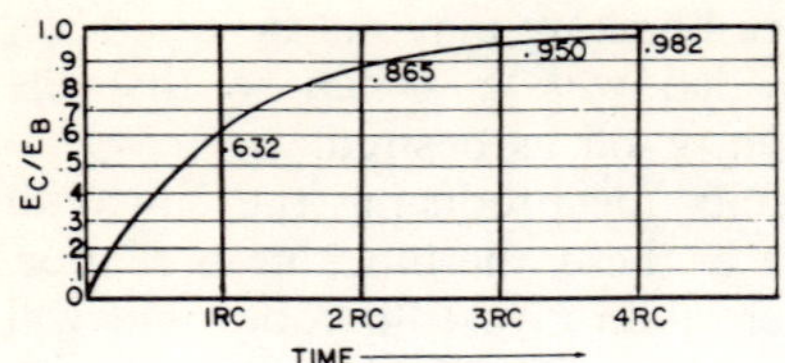

Fig. 3-3. Graph showing the rise in voltage as a capacitor is charged through a resistor.

integrating circuit so that the capacitor charges exponentially, as has been explained. In other words, the charge builds up gradually along a curved path. Then, when the capacitor voltage reaches the firing potential of the four-layer diode, the diode suddenly conducts (switches) and discharges the capacitor rapidly. Next, when the capacitor voltage falls to the cutoff potential of the diode, conduction stops (the diode switches) and the capacitor starts to charge again. In turn, a semisawtooth waveform is generated. Since the amplitude of this sawtooth waveform is usually too small for direct application to the horizontal deflection plates in the crt, most oscilloscopes have one or more stages of horizontal amplification between the sawtooth oscillator and the deflection plates.

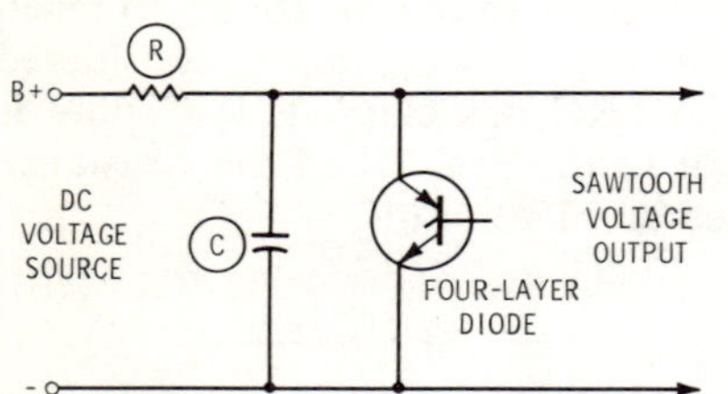

Fig. 3-4. Configuration of a four-layer-diode sawtooth generator.

If the value of R is varied in Fig. 3-4, the time constant of the integrator is thereby changed, and the repetition rate of the sawtooth waveform is also changed. Or, the repetition rate can be changed by varying the value of C. An increase in the time constant lowers the repetition rate, and vice versa. The repetition rate of the sawtooth can also be changed by varying the value of the dc source voltage. If the value of the source voltage is increased, the repetition rate of the sawtooth is thereby increased (the capacitor charges faster). Note that another important characteristic of the oscillator is also changed by increasing the source voltage. With reference to Fig. 3-5, observe that the linearity of the semisawtooth waveform is improved because the high and low switching potentials are then farther down on the total charge curve.

In other words, to increase the linearity of the output waveform and to maintain the same repetition rate, the supply voltage can be increased, thus producing an accompanying increase in the time

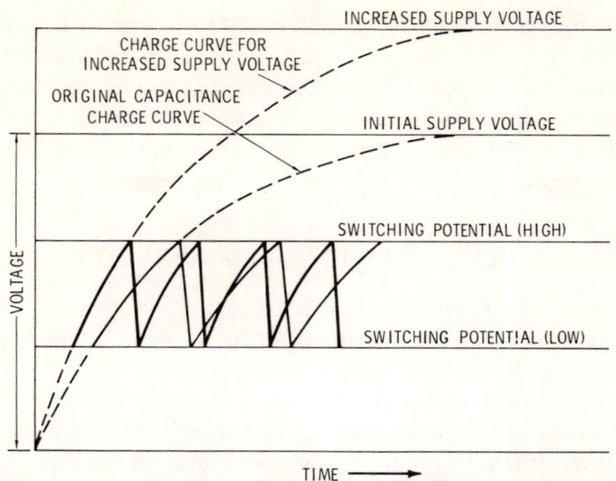

Fig. 3-5. Frequency and linearity changes caused by variation in supply voltage.

constant of the integrating circuit. Next, a wide and continuous range of sawtooth repetition rates must be provided in a horizontal-sweep (time-base) system. In other words, operating controls are provided for adjusting the time constant of the integrating circuit. This is generally accomplished with variable R (potentiometer)

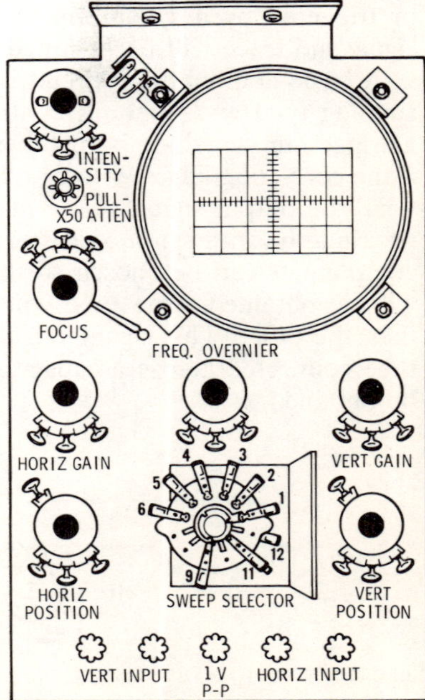

Fig. 3-6. Potentiometer and switch operating controls for an oscilloscope.

control for fine or vernier frequency variation and with half a dozen capacitors with switch control for coarse (step) frequency variation. Fig. 3-6 depicts the potentiometer and switch control facilities for a small service-type oscilloscope. A rotary switch is employed for the sweep selector (step frequency variation). This topic is explained in greater detail in a following chapter.

MULTIVIBRATOR SWEEP CIRCUITS

Some form of multivibrator sweep circuit is almost always utilized as a sawtooth oscillator in modern scopes because the multivibrator can generate the higher sweep rates required. A number of general-purpose oscilloscopes have been designed with sweep rates of several hundred kilohertz. The multivibrator is also widely used as a sweep oscillator in tv receivers, and its design differs very little for the two applications, except for the greater frequency range required in the oscilloscope.

The nature of a multivibrator sweep oscillator is such that it can be easily designed to give either a single sweep, triggered sweeps, or free-running sweeps. A free-running sweep is one operating at its own natural frequency in the absence of a synchronizing signal. A triggered sweep has a natural frequency determined by its circuit components, but each cycle of sweep must be initiated or triggered by a synchronizing signal. In the absence of such a signal, no trace will be obtained with the latter type of sweep.

A basic multivibrator circuit is shown in Fig. 3-7. This is a free-running type and operates continuously without the necessity for a triggering signal. If both tubes have similar characteristics and if the corresponding resistors and capacitors for each tube are identical, the output signal at the plate of tube V2 will be a close approximation to a symmetrical square wave. Values of the different components can be chosen so that a nonsymmetrical square wave will be obtained. One tube will conduct for a much longer time than the other. This signal can then be used to trigger a discharge tube connected across a capacitor, and thus a sawtooth curve will be obtained.

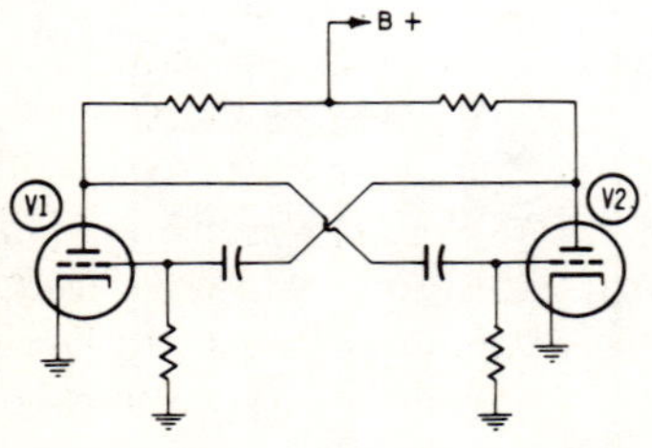

Fig. 3-7. Basic multivibrator circuit.

By further modifications of the circuit, it is possible to eliminate the discharge tube. That function would be performed by the second tube (V2) in addition to its function as part of the multivibrator. Twin-triode tubes are particularly adaptable to multivibrator circuits because they contain in one envelope two triodes of identical characteristics.

A basic transistor multivibrator circuit is shown in Fig. 3-8. This free-running (astable) multivibrator is essentially a nonsinusoidal two-stage amplifier with its output coupled back to its input. As a result, one transistor conducts while the other is cut off, until a point is reached at which the stages reverse their conditions. In other words, the stage which had been conducting cuts off, and the stage that had been cut off conducts. This oscillatory action is used, for example, to produce a square-wave output. By suitable modification, the circuit can be made to produce a sawtooth-wave output. Comparison of Figs. 3-7 and 3-8 shows that the transistor multivibrator is a counterpart of the electron-tube multivibrator.

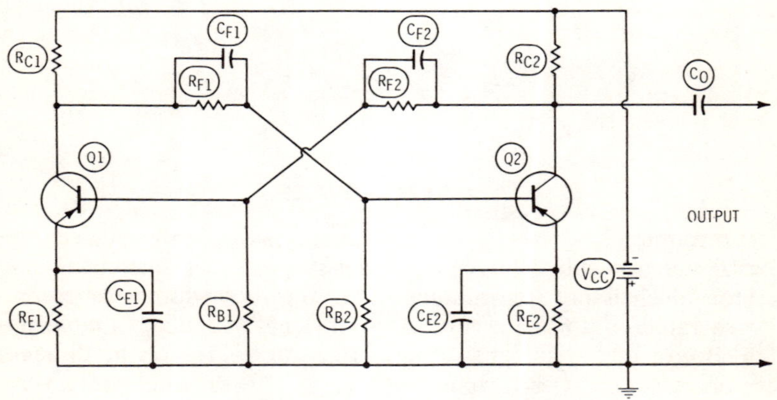

Fig. 3-8. Basic transistor multivibrator circuit.

The foregoing discussion of multivibrator circuit action is summarized by the waveforms depicted in Fig. 3-9 for the circuit of Fig. 3-8. Note that i_{b1} denotes the base current of Q1, V_{b1} denotes the base voltage of Ql, i_{c1} denotes the collector current of Ql, V_{c1} denotes the collector voltage of Ql, i_{b2} denotes the base current of Q2, V_{b2} denotes the base voltage of Q2, i_{c2} denotes the collector current of Q2, and V_{c2} denotes the collector voltage of Q2. Although the collector waveforms are semisquare and have no resemblance to a sawtooth wave, we will find that comparatively simple circuit modifications can provide a collector voltage waveform that is a good approximation to a sawtooth waveform.

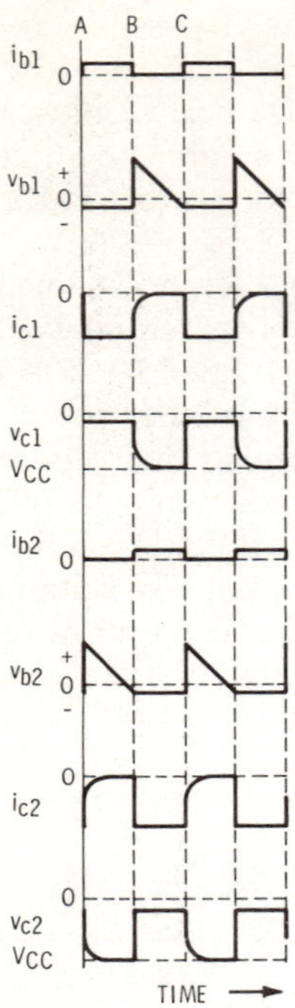

Fig. 3-9. Voltage and current waveforms in a basic transistor multivibrator circuit.

TRIGGERED SWEEP

Conventional scopes use a free-running sawtooth oscillator for horizontal deflection. *Triggered-sweep* scopes use a sawtooth generator which is not free-running. Therefore a sawtooth waveform is generated only when a vertical-input signal is applied. The leading edge of the input signal (Fig. 3-10) triggers the sawtooth generator, and one sweep excursion occurs. The sawtooth generator then remains inactive until the next leading edge arrives. This means that before the leading edge of a vertical input signal arrives, you see only a spot on the screen, as illustrated in Fig. 3-11A. When the leading edge arrives, the beam deflects horizontally, as seen in Fig. 3-11B. In turn, a waveform can be greatly expanded

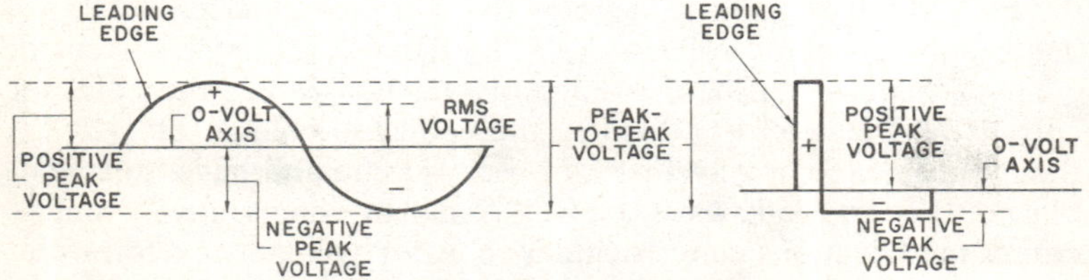

Fig. 3-10. Leading edges in sine and pulse signal waveforms.

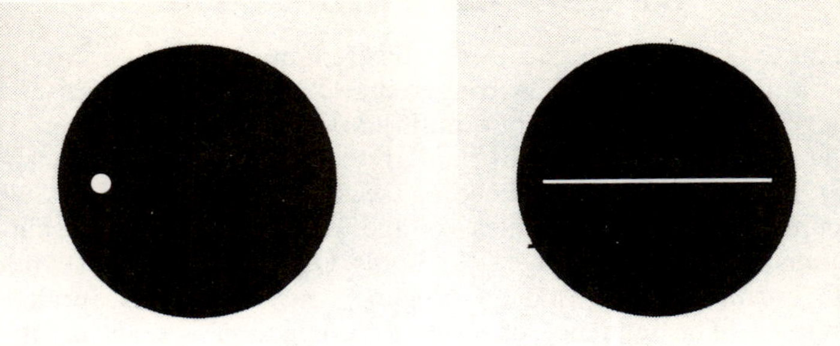

(A) Spot resting; not triggered.

(B) Scope triggered.

Fig. 3-11. Trigger control of the cathode-ray tube beam.

merely by advancing the horizontal sweep-rate control (Fig. 3-12). Trigger action is obtained by biasing at least one of the tubes or transistors in a multivibrator beyond cutoff. For example, in a cathode-coupled multivibrator a variable common-cathode bias control may be provided to set the triggering level.

Some oscilloscopes designed for industrial applications have provision for very slow horizontal sweep. A binding post may be provided on the front panel, marked "external capacitor." When a capacitor is connected between the binding post and ground, the time constant of sweep oscillator is increased. In turn, the horizontal-deflection rate slows down. The scope manufacturer often supplies a chart that shows the sweep rate versus the external capacitance value.

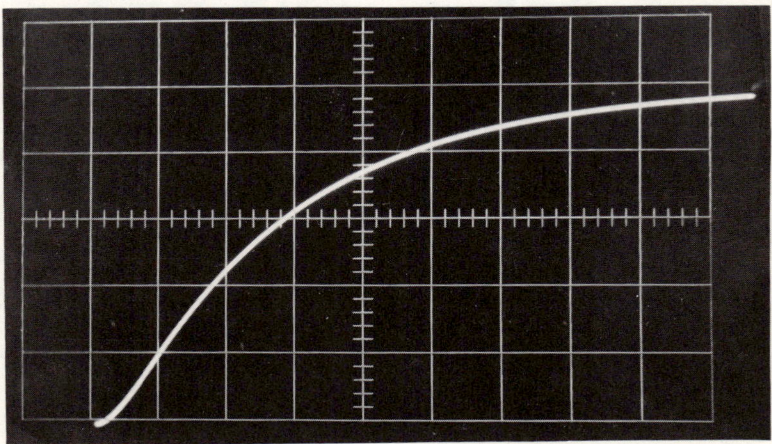

Fig. 3-12. Expansion of the leading edge of a waveform.

TRANSISTORIZED SWEEP SECTION

You will find transistorized triggered-sweep configurations in most modern lab scopes. A triggered sweep, or time base, uses the basic monostable (one-shot) multivibrator circuit shown in Fig. 3-13A. This configuration is similar to the multivibrator arrangement described in the preceding section, "Multivibrator Sweep Circuits," except that the bias voltage V_{BB} holds Q1 below cutoff, whereas the supply voltage V_{CC} holds Q2 in saturation (conduction) during the time that no trigger (sync) voltage is applied to Q1 via C_c. The sequence of multivibrator action is seen in Fig. 3-13B. When a negative sync voltage (trigger pulse) is applied to

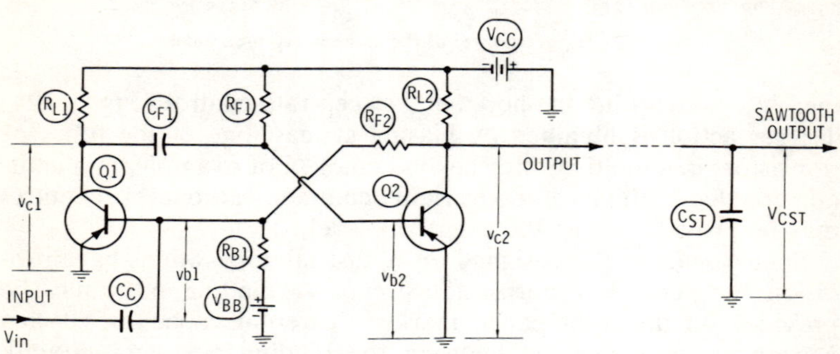

(A) Multivibrator configuration.

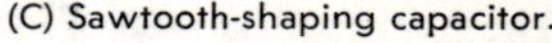

(C) Sawtooth-shaping capacitor.

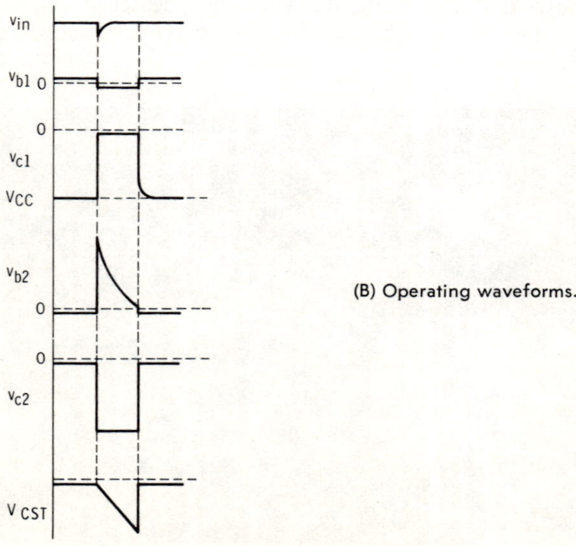

(B) Operating waveforms.

Fig. 3-13. Basic monostable (one-shot) multivibrator.

the base of Q1 (see v_{in}), the transistor is momentarily driven into conduction. In turn, the trigger pulse is amplified and applied to the base of Q2 in positive polarity. Thus, Q2 is cut off (see v_{b2}) and Q1, in turn, is held in conduction until the charge on C_{F1} decays through R_{F1}. At this point, Q2 goes back into saturation and Q_1 goes back below cutoff. The result is a single square output pulse (see v_{c2}). The multivibrator then rests in its quiescent state until another trigger pulse is applied to the base of Q1.

Before this square output pulse can be used to deflect the crt beam, it must be changed into a sawtooth form. This is the function of C_{st} in Fig. 3-13C. Note that during the time that Q2 is cut off, C_{st} charges through R_{L2} and forms one sawtooth excursion. Next, when Q2 suddenly goes into saturation again, C_{st} discharges rapidly through Q2. The resulting sawtooth wave, V_{cst}, is in turn applied to the horizontal-deflection plates in the scope, and produces one excursion of the triggered-sweep action. It is evident that the sweep speed depends on the value of C_{st} (time constant of the charging circuit). In practice, it is necessary to elaborate the basic time base somewhat. First, if an arbitrary vertical-signal waveform were used to trigger the multivibrator, the resulting trigger action would often be erratic. Therefore, a waveshaping section must be employed which produces a standard trigger pulse each time that the leading edge of the vertical-signal waveform appears. Second, control facilities must be provided whereby this standard trigger pulse is produced when the vertical-signal leading edge is positive-going or when it is negative-going.

Fig. 3-14 shows a basic waveshaping section which is used to trigger the monostable multivibrator. This waveshaping arrangement is called a Schmitt trigger circuit. It is essentially a bistable multivibrator. In other words, if Q1 is cut off, Q2 will be saturated, and the circuit will remain in this state until the base voltage on Q1 is changed to bring Q1 into conduction. Thereupon, Q2 is driven into cutoff, and the circuit will remain in this reversed state until the base voltage on Q1 is again changed to cut off Q2. In turn, Q2 simultaneously goes into saturation. This circuit action is a result of the common-emitter resistor, R_E, and the bias source, V_{EE}. Input and output waveforms are shown in Fig. 3-14B, for the case of a sine-wave input. Observe that the output is a square wave. No matter what the input waveform may be, the output will remain a square waveform. Thus, uniform triggering of the monostable multivibrator is ensured. Note that when the leading edge of the output waveform in Fig. 3-14B is applied to the input lead of the multivibrator in Fig. 3-13A, the square waveform becomes differentiated by C_c and R_{B1}, thereby forming the trigger pulse, v_{in}, depicted in Fig. 3-13B.

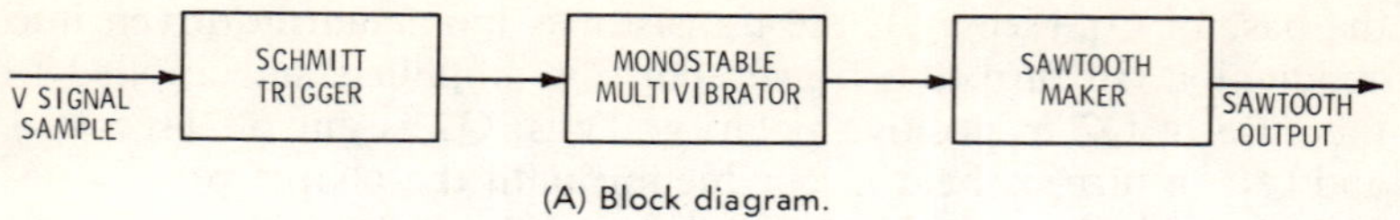

(A) Block diagram.

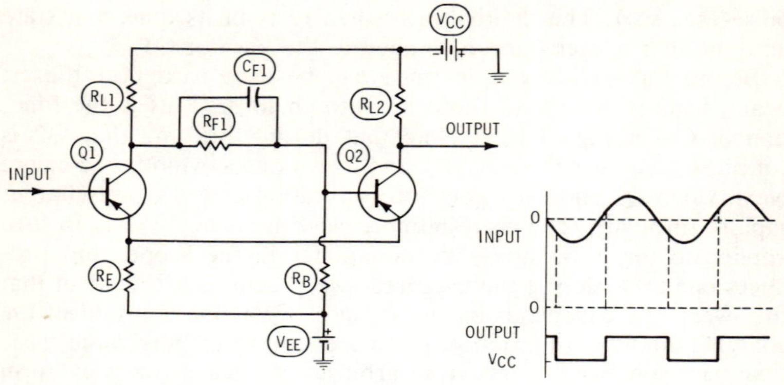

(B) Schmitt trigger configuration.

Fig. 3-14. Monostable multivibrator preceded by a Schmitt trigger circuit.

We observe that triggering of the time base can occur only on the negative-going excursion of the input waveform in Fig. 3-14B. Of course, this is not always desirable—we might wish to display a positive-going waveform instead. Therefore, a triggered-sweep scope provides facilities for either positive or negative triggering. For this purpose, the Schmitt trigger in Fig. 3-14A is preceded by a phase inverter. A phase inverter effectively turns the input waveform "upside down" so that its negative-going leading edge is changed into a positive-going leading edge. Similarly, the positive-going leading edge of the input waveform is changed into a neg-

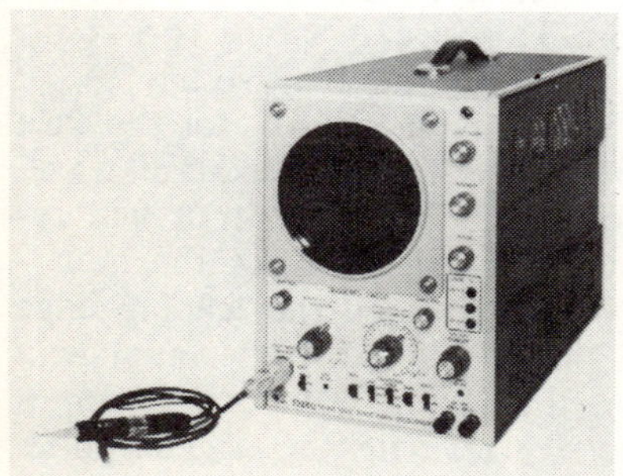

Courtesy EICO

Fig. 3-15. A typical service-type triggered-sweep oscilloscope.

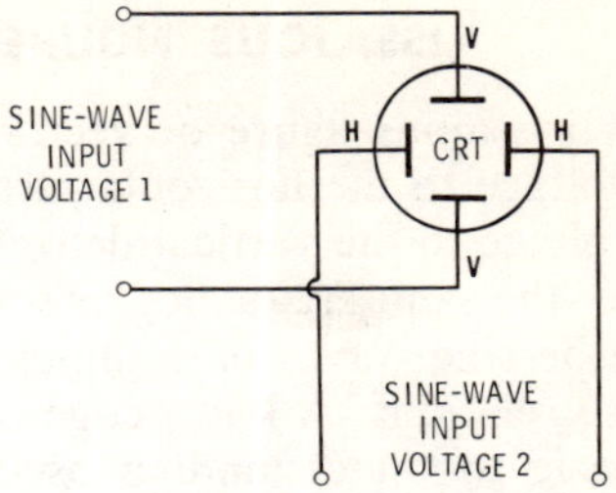

SINE-WAVE INPUT VOLTAGE 1

V

H

CRT

H

V

SINE-WAVE INPUT VOLTAGE 2

Fig. 3-16. Sine-wave horizontal- and vertical-deflecting voltages produce Lissajous figures.

ative-going leading edge. This change permits switch control of negative-going triggering or positive-going triggering. Still other elaborations of the basic triggered time base are provided in modern scopes, as explained in greater detail in the next chapter. A standard service-type triggered-sweep oscilloscope is pictured in Fig. 3-15.

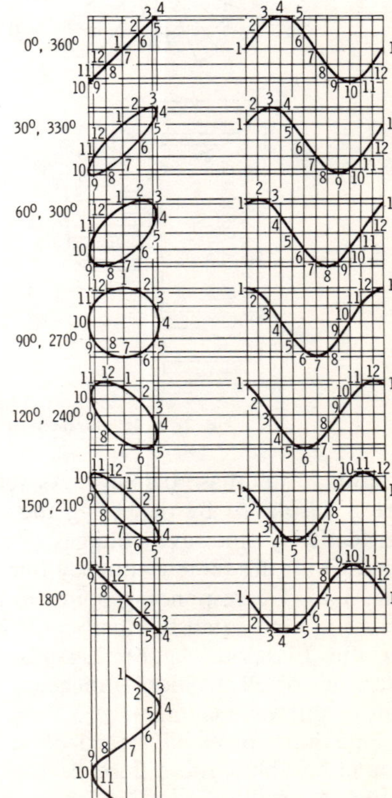

Fig. 3-17. Basic Lissajous figures for sine waves with the same frequency but different phases.

LISSAJOUS FIGURES AND VECTORGRAMS

A Lissajous figure or vectorgram is displayed by applying one ac voltage to the horizontal-deflection plates and applying another ac voltage to the vertical-deflection plates. These two voltages have either the same frequency or some integral frequency relation. They may be sine waves or complex waves, and they may have various phase relations. A knowledge of Lissajous figures and vectorgrams is basic to understanding oscilloscopes because Lissajous figures and vectorgrams are another type of display in comparison to con-

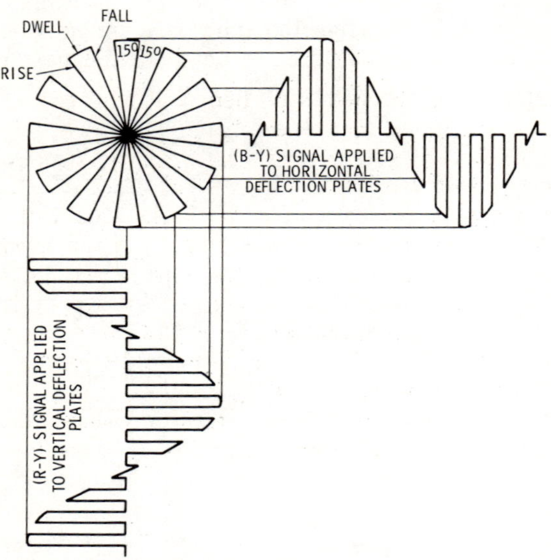

Fig. 3-18. Development of an ideal vectorgram.

ventional time-base displays. As seen in Fig. 3-16, a Lissajous figure is displayed by deflecting the crt beam both horizontally and vertically by sinewave voltages. Various Lissajous figures for sine waves with the same frequency but different phases are depicted in Fig. 3-17. Corresponding points in time on the V and H input sine waves are indicated by numbers, with corresponding points in time on the Lissajous figure. Lissajous figures are used in various branches of electronics to measure the phase difference between a pair of sinewave voltages.

You have probably heard of vectorscopes, and you may have wondered about the difference between a vectorscope and a conventional oscilloscope. The fact is that there is very little difference,

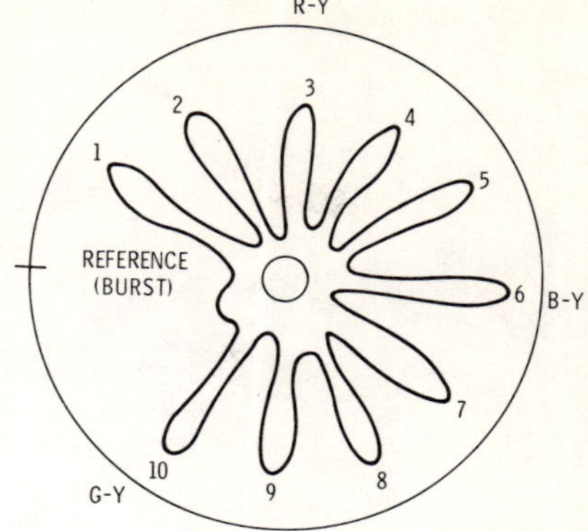

Fig. 3-19. Typical vectorgram obtained in servicing procedures.

and sometimes no difference at all between these two instruments. Just as Lissajous figures can be displayed with an ordinary oscilloscope, so can vectorgrams in nearly all cases. First, it is helpful to observe how a vectorgram display is obtained. This procedure involves the use of a standard color-bar generator with a color-tv receiver. Without going into unnecessary details, modified sine-wave voltages are produced by the color receiver. These voltages are called R-Y and B-Y waveforms, and are applied to the crt deflection plates, as was shown in Fig. 3-16. The result is a display of a vectorgram, depicted in ideal form in Fig. 3-18. A typical vectorgram obtained in practice is pictured in Fig. 3-19. Interested readers are referred to specialized books on color-tv servicing for further information on this topic.

Chapter 4

Synchronization

Synchronization can be defined as the timing of two or more events so that they will occur simultaneously or in step with each other. As applied to oscilloscopes, such timing would be between the signal to be viewed and the sweep system of the oscilloscope. In the majority of cases, the technician will be observing a signal that has some regularly recurring peak or dip in its amplitude. In other words, it is made up of cycles that repeat regularly and can therefore be synchronized with the trace of the oscilloscope as the trace sweeps across the oscilloscope screen.

When an oscilloscope has been properly synchronized with a signal, this signal will appear to be stationary on the screen, and one or more cycles of the signal can be seen. If the oscilloscope is slightly out of synchronization, the waveform will appear to move slowly across the screen, either to the right or to the left. The waveform can be made to "stand still" even without a sync signal if the operator carefully adjusts the fine frequency control. It will not remain stationary very long, though, because of the tendency of the sweep oscillator to wander in frequency. The situation is quite similar to that of a television receiver which has lost the sync signal: The receiver operator varies the frequency of the sweep oscillator by adjusting the hold control, and when the oscillator frequency agrees with the sweep frequency of the tv signal, the picture is held stationary on the screen, but only as long as the operator is willing to keep adjusting the hold control.

Certain types of signal will be more difficult to synchronize than others. These include signals having little information of a recurring or repeating nature and signals having few pronounced peaks or dips. As the frequencies of both the signal being viewed and the oscilloscope sweep are increased, synchronization also becomes more difficult. The reason for these difficulties will become more apparent later when we consider the process of synchronization.

If a signal does not repeat at regular intervals, it is called a transient or nonrecurrent waveform. Even if a signal repeats at regular intervals, it may approximate a transient in case there is a long waiting interval between the repetitions. Sometimes there may be many waveforms of no interest that occur between the repetitions of the desired waveform. As an illustration, a vertical-interval test signal is transmitted in each vertical-blanking interval of the composite color-tv signal. The foregoing waveforms cannot be displayed with a scope that has free-running sweep. On the other hand, such waveforms can be displayed with a scope that has triggered sweeps. It is sometimes advantageous to utilize a long-persistence crt, particularly when the signal is of the "one-shot" type and is not repeated at any time. Note that highly sophisticated lab-type scopes which have a screen-pattern storage function are also available. In other words, a "one-shot" transient can be maintained visibly on the screen for hours, if desired.

Various general-purpose scopes provide a choice of recurrent-sweep mode and trigger-sweep-mode operation. This is called a dual-mode sweep function. A scope with this feature is pictured in Fig. 4-1. In its recurrent-sweep mode, it is a conventional continuous-sweep type of oscilloscope. The sweep oscillator is adjustable up to 1 MHz in six ranges, permitting lock-in of high-frequency signals. The sweep ranges are marked not only with frequency indications, but also with time-base references (milliseconds or microseconds per centimeter). Preset tv V and H sweep positions are provided for convenience in tv servicing. A 60-Hz sine-wave "Line" horizontal-deflection function is also provided, with adjustable phase, for use in sweep-alignment procedures. This topic is explained further in a following chapter.

SYNCHRONIZATION OF MULTIVIBRATORS

Synchronization of multivibrator sweeps is fundamental in oscilloscope operation. A typical multivibrator circuit uses both sections of a twin triode with the signal from each plate coupled to the grid of the other section. Each section is alternately cut off while the other conducts. By proper choice of circuit constants, an unbalanced multivibrator is obtained with one section remaining cut off for a much greater time than its conduction period. The nonconduction period is used as the charging time for a capacitor, thus developing a sweep voltage, and the short conduction period is used to discharge the capacitor, thus developing the retrace. A synchronizing signal can be used to influence the time at which the sweep section changes to a conductive or nonconductive state. Such a signal is usually introduced at the grid of one of the sections.

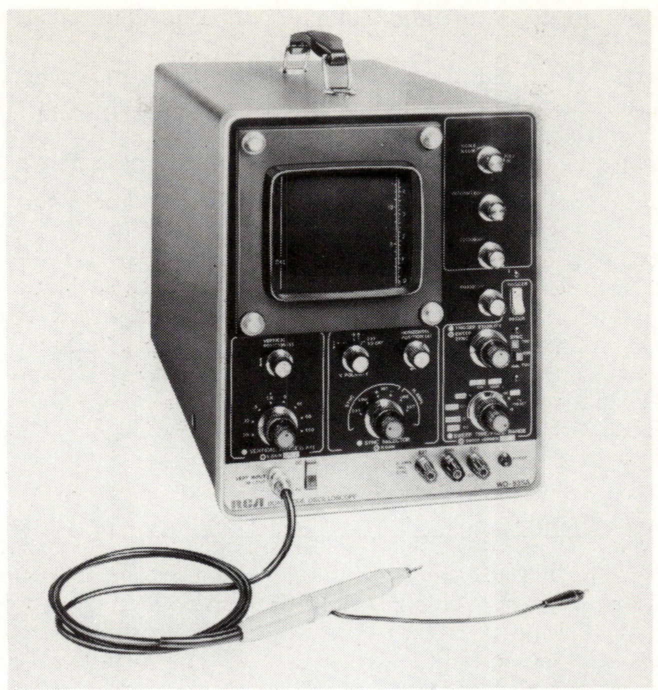

Courtesy RCA Corp.

Fig. 4-1. A general-purpose oscilloscope with a dual-mode sweep function.

The waveforms obtained at various points in a typical multivibrator circuit can be found in many textbooks. Fig. 4-2 shows one such waveform, the waveform developed at the grid of one section of a balanced multivibrator. That is, it is one in which the conduction and nonconduction periods of the sections are equal.

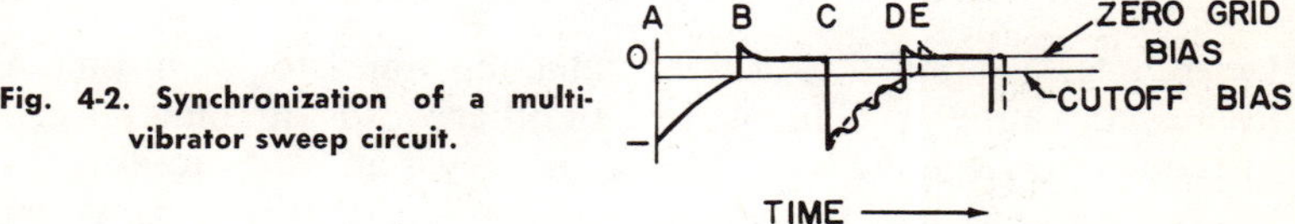

Fig. 4-2. Synchronization of a multivibrator sweep circuit.

At time A, the tube section is cut off by a comparatively large negative bias on its grid. Between times A and B, the RC networks affecting the section gradually lose their negative charge, and the grid bias rises in a positive direction until it reaches the cutoff value at B, where the tube section suddenly changes to a conductive state and remains so until time C. At time C, we have shown how a sync signal introduced at the grid will affect the conduction point on the next cycle.

Next, it is helpful to consider the use of pulses to synchronize a transistor multivibrator circuit. Fig. 4-3 shows the effect of positive pulses on the multivibrator base waveform. Note that when a positive pulse of insufficient amplitude to drive the base above cutoff

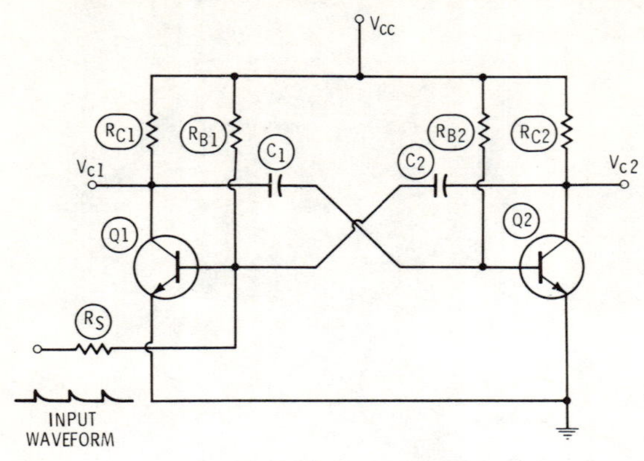

(A) Configuration.

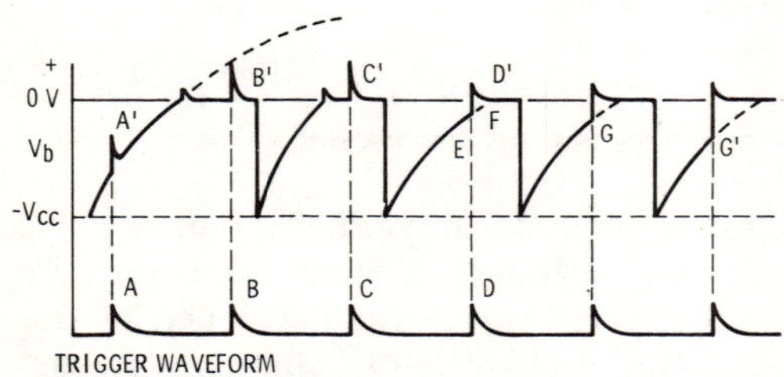

(B) Base waveform with positive sync pulses.

Fig. 4-3. Synchronization of a transistor multivibrator.

(zero volts) is applied to a nonconducting transistor at instant A, it does not cause switching action. In other words, its only effect is to reduce the negative bias slightly, as shown at A′. Note also that when a positive pulse is applied to a transistor that is already conducting (points B or C), the pulse serves merely to increase the base voltage momentarily. That is, the pulse has no effect on mutivibrator action at this instant. Observe that with the exceptions of the variations in the v_b waveform at times A′, B′, and C′, the multivibrator is essentially free-running. However, if the positive pulse is applied at instant D, it has sufficient amplitude to overcome the negative voltage on the base of Q1, and it drives the base above zero volts. Accordingly, the period of the multivibrator is shortened by the interval, EF.

We will recognize that for proper synchronization, the natural period of the multivibrator in its free-running state must be greater than the time interval between sync pulses. Under this condition, the positive trigger pulses cause the switching action to occur earlier in the cycle than it would occur in the free-running state. Hence, the transistor conducts at E, G, and G′ in Fig. 4-3. Otherwise, in the absence of sync pulses, the transistor would have conducted later in each instance. In this manner, synchronizing action forces the multivibrator frequency to become the same as the repetition rate of the sync pulses. Of course, the multivibrator may be synchronized to a submultiple of the trigger frequency, if both frequencies are such that every second, third, fourth, etc., sync pulse occurs at the correct time to drive the base voltage of the nonconducting transistor above cutoff.

TRIGGERED-SWEEP SYNCHRONIZATION

Synchronization of a triggered-sweep scope is slightly more complex than synchronization of an ordinary service-type scope. In a typical arrangement (Fig. 4-4), a sync phase splitter is followed by a *sync-level* control. The sync-level control determines the amplitude of the sync signal fed to the sync trigger generator. If the level is set too low, the trigger generator will not operate. On the other hand, if the level is set too high, part of the leading edge may be lost in the displayed waveform.

The sync trigger generator is followed by the *stability* control. This control must be set within the limits of gate-generator operation; otherwise the pattern may jitter or "run" on the screen. The circuits following the gate generator form the horizontal-deflection sawtooth. Sweep speed is adjustable by means of the *sweep-rate* control and its *multiplier,* and the *sweep-vernier* control. Individual sweep speeds are accurately calibrated in microseconds per centimeter, milliseconds per centimeter, or seconds per centimeter. High accuracy of sweep speed is obtained because the sawtooth discharge tube is isolated from the sync signal by the gate generator.

Many triggered-sweep scopes provide an *automatic* switch position which converts the triggered-sweep section into a free-running sweep. This feature is helpful for the inexperienced user, because he can set the centering controls and vertical-gain controls for a satisfactory pattern size and position before he concerns himself with the triggered-sweep controls.

There has been a trend to use transistorized sync sections in modern lab-type scopes as well as to use completely transistored designs. A *hybrid* configuration uses tubes in the input stage and transistors elsewhere. The tubes serve two functions: first, they provide comparatively high input impedance; second, tubes can be driven into saturation without overload damage. Thus, the tubes serve as protective buffers between the sync-input signal source and the trigger-amplifier transistors. The function of the Schmitt trigger generator is to supply a uniform 75-volt gate pulse to the sweep section, regardless of the shape or amplitude of the incoming sync-input signal.

With reference to Fig. 4-4, the sync-level control with its positive and negative ranges permits the sweep circuit to be triggered at any point on the leading or trailing edge of an input waveform.

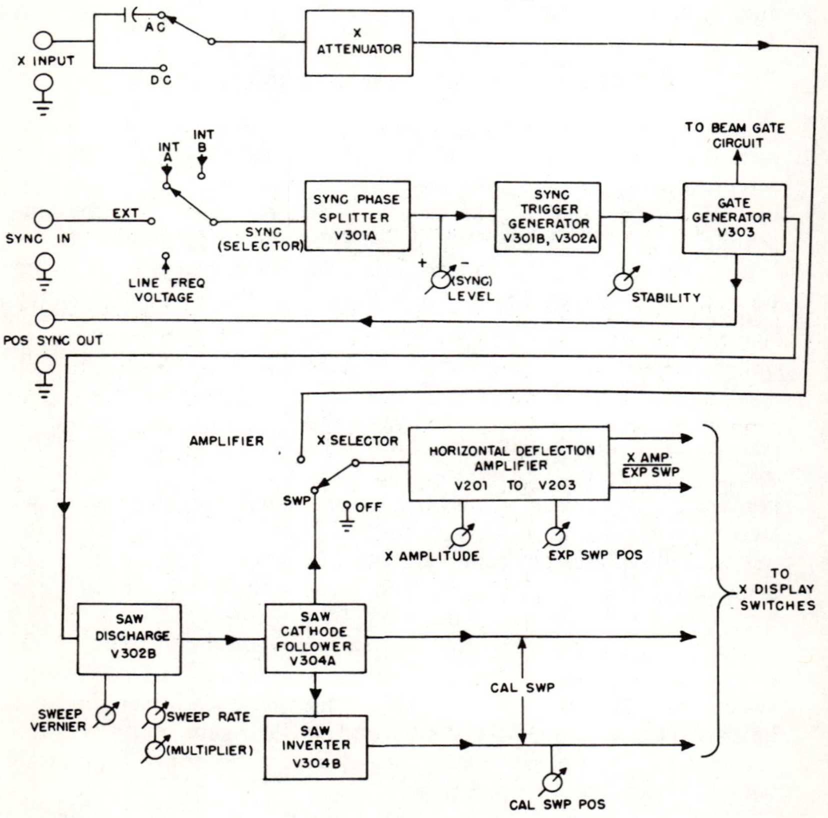

Fig. 4-4. Block diagram of a triggered-sweep system.

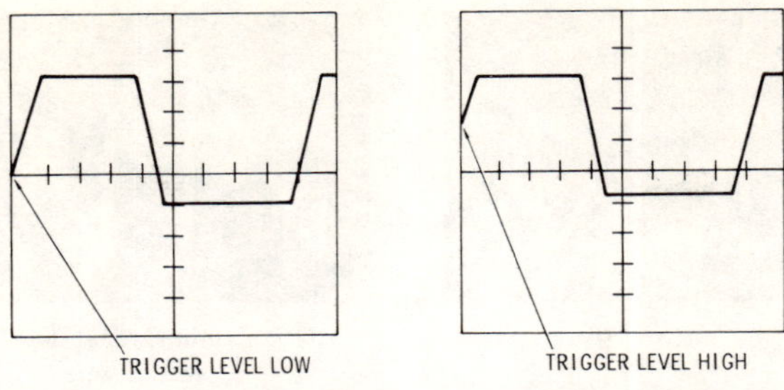

(A) Positive slope of input signal used as trigger (rising interval).

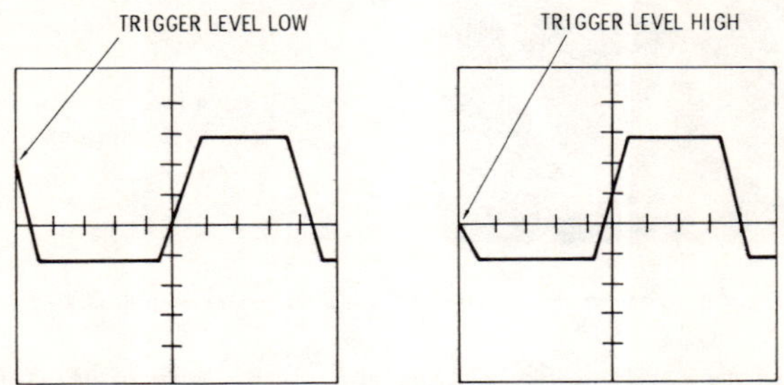

(B) Negative slope of input signal used as trigger (falling interval).

Fig. 4-5. Trigger-sweep controls permit triggering at any point on a leading or trailing edge.

Typical high and low trigger points on leading and trailing edges of a trapezoidal waveform are depicted in Fig. 4-5. It is also possible to trigger the sweep circuit at any point along the flat top of the waveform. However, the internal-sync function of the scope cannot be used in this case. Instead, the external-sync setting in Fig. 4-4 must be utilized. In turn, a suitable sync signal is applied to the sync-input terminals. This sync signal is typically obtained from a pulse generator which, in turn, is locked to the vertical-input signal. Effectively, a delayed sync pulse is applied to the sync phase splitter in order to trigger the sweep circuit at some point along the flat top of the waveform being displayed. The exact point at which the sweep circuit is triggered depends on the repetition rate to which the pulse generator is set.

OVERSYNCHRONIZATION

Both types of sweep oscillators are subject to oversynchronization if too much sync signal is applied, and the effects on the resulting waveforms are similar. Fig. 4-6A and B show actual waveforms photographed at points in a multivibrator sweep circuit of

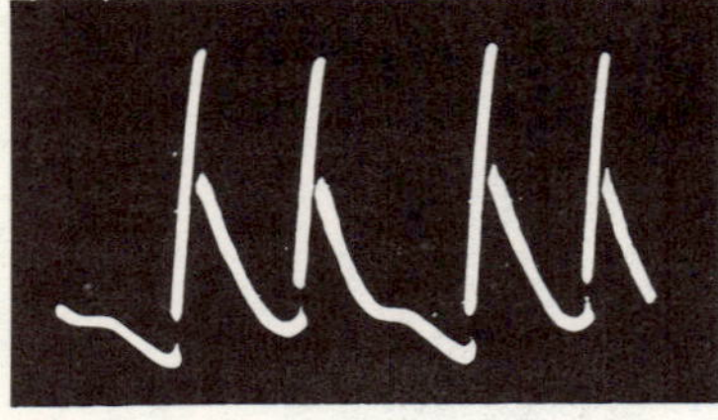

(A) Signal at the plate of first section.

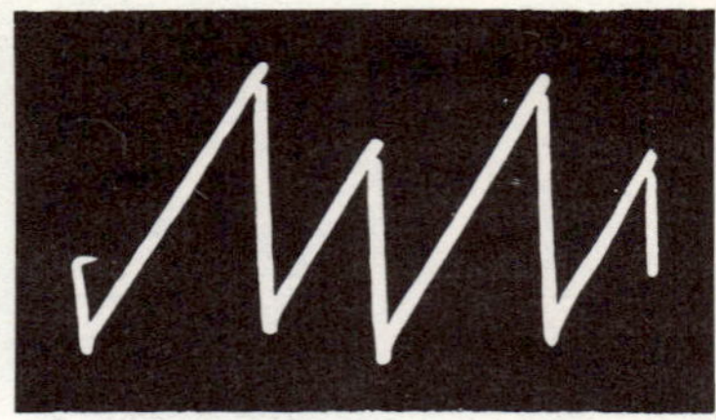

(B) Sawtooth signal developed at second section.

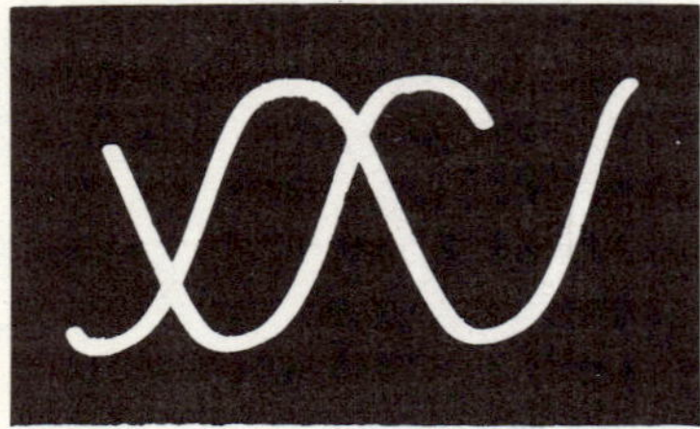

(C) Waveform seen on oscilloscope screen.

Fig. 4-6. Oversynchronization of a multivibrator sweep circuit.

an oscilloscope. Fig. 4-6A shows the signal obtained at the plate of the first section, and Fig. 4-6B, the signal at the output of the discharge section. Fig. 4-6C illustrates the waveform actually displayed on the screen of the oscilloscope. Fig. 4-6A shows that one cycle of sweep contains parts of three cycles of sine-wave signal and that the other cycle of sweep contains parts of two cycles of signal. Fig. 4-6B depicts that the sweeps travel at a constant rate but that alternate cycles are of different lengths. Each cycle of sweep and retrace can be seen to correspond to certain portions of the waveform of Fig. 4-6C.

If oversynchronization is carried to extremes, the sawtooth waveform of Fig. 4-6B may even degenerate into one large cycle followed by two or three very small ones. In such a case the waveform on the screen would also be greatly distorted. Some manufacturers have designed circuits to lessen or eliminate the possibility of oversynchronization. One method is to use a limiter stage ahead of the point of injection of the sync signal. In oscilloscopes that do not have provision for limiting oversynchronization effects, it is sometimes found that a change in the vertical amplitude setting will affect the sweep operation, causing the waveform to fall either into or out of sync. These are usually cases where the sync takeoff point follows the vertical amplitude control. Therefore, when the vertical amplitude is changed, the amplitude of the sync signal changes with it, with the result that synchronization may be affected.

For simplicity the sync signals shown in the preceding examples have been sine-wave signals. In cases where the sync signal is taken as a part of the signal in the vertical amplifiers of the oscilloscope, it can take on any form, depending on the waveform being viewed. Generally, stable synchronization will be more easily attained with signals of a peaked or sharply pulsed nature rather than with those of a more even nature. Sometimes a signal may have more peaks at its negative extreme than at its positive extreme or vice versa. A sync polarity-reversal switch on the oscilloscope may help the operator obtain stable synchronization in these cases. If such a switch is not included, the same effect can sometimes be obtained if the signal to be observed is taken from a point having a signal 180 degrees out of phase with respect to the signal at the first point. The signals between two successive stages in an amplifier usually have this phase reversal.

Chapter 5

Amplifiers

Practically all modern oscilloscopes contain amplifiers to increase the signal amplitude before it is applied to the vertical or horizontal deflection plates of the cathode-ray tube. These same oscilloscopes usually have provision for making direct connection to the deflection plates without benefit of the amplifiers; however, a signal of comparatively high amplitude is required to obtain a useful deflection when making a direct connection to the deflection plates. Amplifiers must therefore be used if low-amplitude signals are to be observed.

For example, one model of oscilloscope is quoted as having a deflection sensitivity of 15 volts rms per inch at the vertical deflection plates. This means a 15-volt rms signal applied to the plates will produce a waveform one inch high. Another model is quoted as having deflection sensitivities of 13 and 17 volts rms per inch, respectively, for the vertical and horizontal deflection plates. The service technician will, in most cases, be dealing with signals having amplitudes much lower than these values. A 15-millivolt signal in the first example would produce a deflection of 1/1000 inch, certainly of no use to the technician. An amplifier with a voltage gain of 1000 will bring the deflection up to 1 inch—a usable deflection, though perhaps not ideal.

A direct connection to the deflection plates may prove advantageous in certain cases even though a comparatively large signal is required. When the oscilloscope is used to make exacting phase comparisons, direct connection to both sets of deflection plates will avoid phase distortion that might otherwise be caused by phase differences between the horizontal and vertical amplifiers. The frequency response will also be improved by direct connections if the response of the oscilloscope amplifiers extends to only a few hundred kilohertz.

The foregoing paragraphs naturally lead to a consideration of two important characteristics of an oscilloscope: the sensitivity and the frequency response of its amplifiers. The service technician is concerned with these two characteristics because they define to a great extent the limits of the usefulness of an oscilloscope. The sensitivity determines the minimum amount of signal, and the frequency response determines the range of frequencies that can be viewed on the oscilloscope. Using a detector probe will indirectly extend the useful range of the oscilloscope to higher frequencies.

At present, general-purpose oscilloscopes having vertical sensitivities of 10, 15, or 20 millivolts rms per inch are on the market. A number of such oscilloscopes also have a frequency response usable up to 4.5 megahertz and beyond. The sensitivity of an amplifier can be increased by adding a number of stages; but since this increase in sensitivity must also be accompanied by a wideband response, the design of such an amplifier becomes more difficult. Most oscilloscopes presently incorporate wideband, push-pull amplifiers. The wideband response for these amplifiers is accomplished in much the same manner as for the video amplifier in a tv receiver —by means of series and shunt peaking, plate loads of low value, and (in some cases) feedback circuits. Sometimes, portions of these circuits are controlled by a switch on the front or rear panel so that the operator can chose between operation at a narrow bandwidth with high sensitivity and operation at a wide bandwidth with medium sensitivity.

Some lab-type oscilloscopes have vertical amplifiers with very considerable bandwidth. For example, one design provides response up to 3900 MHz. To obtain this extreme bandwidth, a sampling technique is used, as pictured in Fig. 5-1. Conventional oscilloscopes are limited in bandwidth to frequencies in the megahertz region. However, sampling oscilloscopes have response into the gigahertz region. This design samples the input waveform and then reconstructs the waveform for display from the samples taken dur-

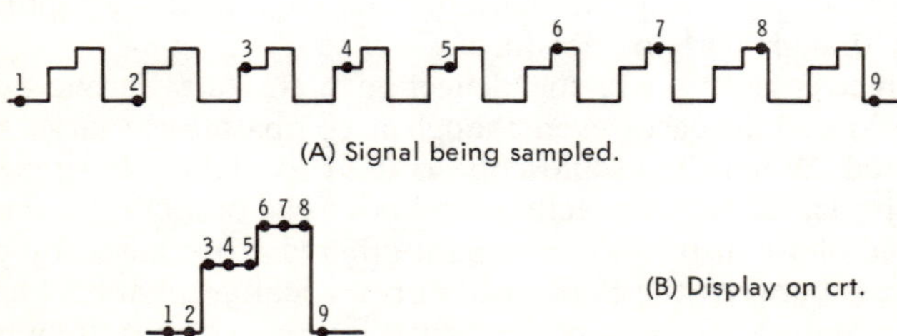

Fig. 5-1. A sampling oscilloscope reconstructs the input waveform by taking successive samples

ing many recurrences of the input waveform. Therefore, a sampling oscilloscope cannot be used to display one-shot transients. To reconstruct a waveform, the sampling pulse "turns on" the sampling circuit for a brief instant, and the amplitude at this point is displayed by the electron beam in the crt. After a brief pause, the following cycle of the input waveform is again sampled, and the amplitude at this next point is displayed by the crt beam which has moved slightly to the right in the meantime. As many as 1000 samples may be processed to reconstruct the input waveform for display on the crt.

HIGH-FREQUENCY RESPONSE

Fig. 5-2 shows an example of the use of series- and shunt-peaking coils to extend the high-frequency response of the vertical amplifier of an oscilloscope. A series-peaking coil is connected between the output of one stage and the input of the next stage. A shunt-peaking coil is connected between the output of a stage and the power supply. For example, L2, L3, L4, and L5 are series-peaking coils; L1 is a shunt-peaking coil. Peaking coils extend the high-frequency response of an amplifier by resonating with the residual circuit capacitances at the high-frequency end of the band. Peaking coils must have correct inductance values to provide a reasonably flat frequency response, as seen in Fig. 5-3. It is undesirable to use overcompensation because the amplifier will develop overshoot and ringing distortion in the displayed waveforms.

Another type of high-frequency compensation is also used in the configuration of Fig. 5-2. Note that R16 is partially bypassed by C8. In other words, a 470-pF capacitor has a reactance of approximately 220 ohms at 1.5 MHz, and a reactance of about 70 ohms at 5 MHz. Therefore, the negative feedback produced by R16 decreases progressively in the range from 1 to 5 MHz, thereby providing greater gain at higher frequencies. Fig. 5-4 shows basic shunt, series, and series-shunt compensation circuits for transistor stages. The residual circuit capacitances are indicated at C_o (output capacitance) and C_i (input capacitance). At high frequencies, L_1 becomes parallel-resonant with C_o and C_i, thereby increasing the effective value of the collector load impedance and, in turn, increasing the stage gain. Again, at high frequencies, L_2 becomes series-resonant with C_o and C_1, resulting in a voltage magnification across L_2 equal to its Q value.

Vertical amplifiers in lab-type scopes with extended high-frequency response often utilize a distributed-amplifier configuration, as shown in Fig. 5-5. This arrangement uses many very small peaking coils in up to 30 stages of amplification. The effect of the small

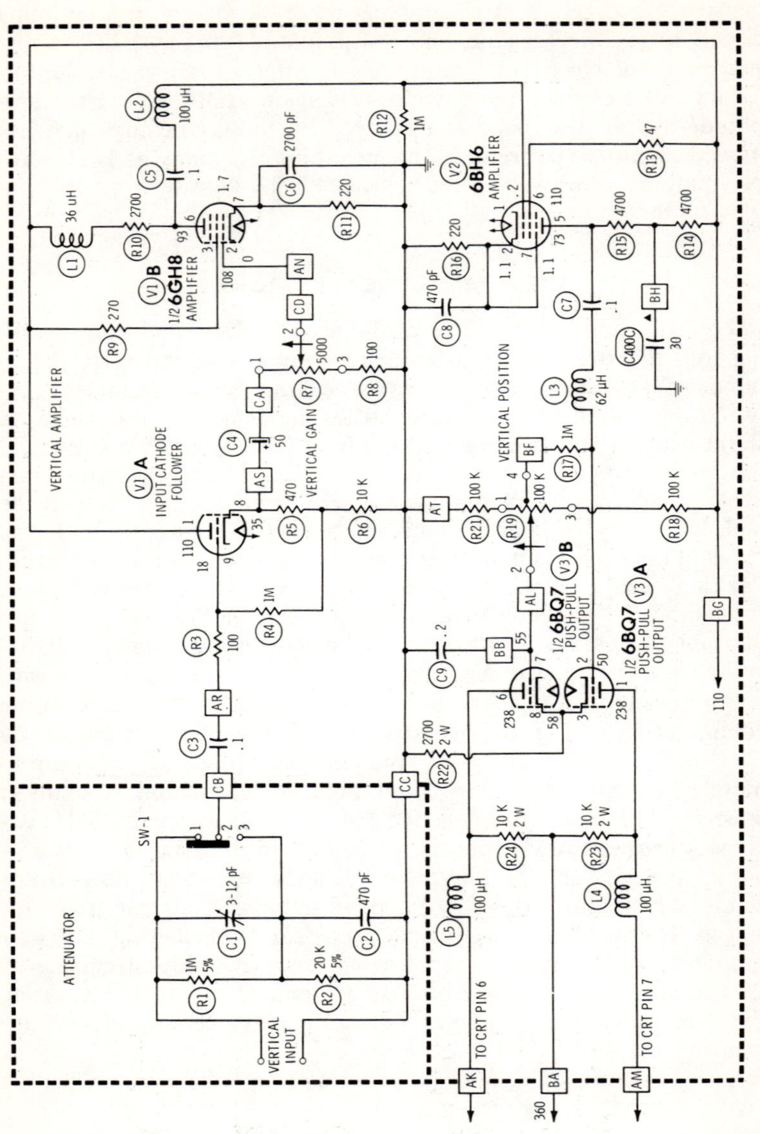

Courtesy Heath Co.

Fig. 5-2. Vertical-amplifier configuration with series and shunt peaking.

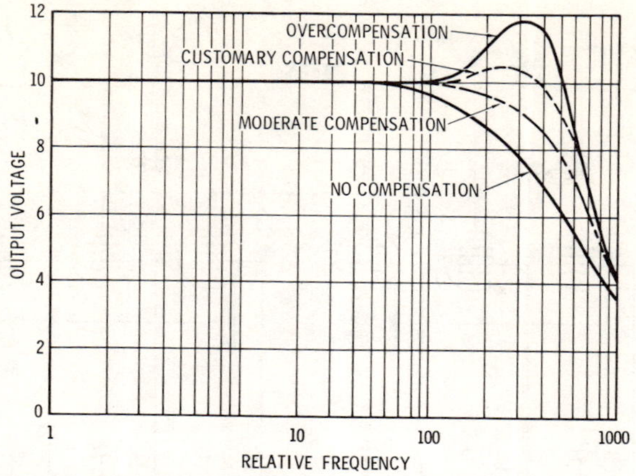

Fig. 5-3. Vertical-amplifier frequency-response curves for various values of peaking-coil inductance.

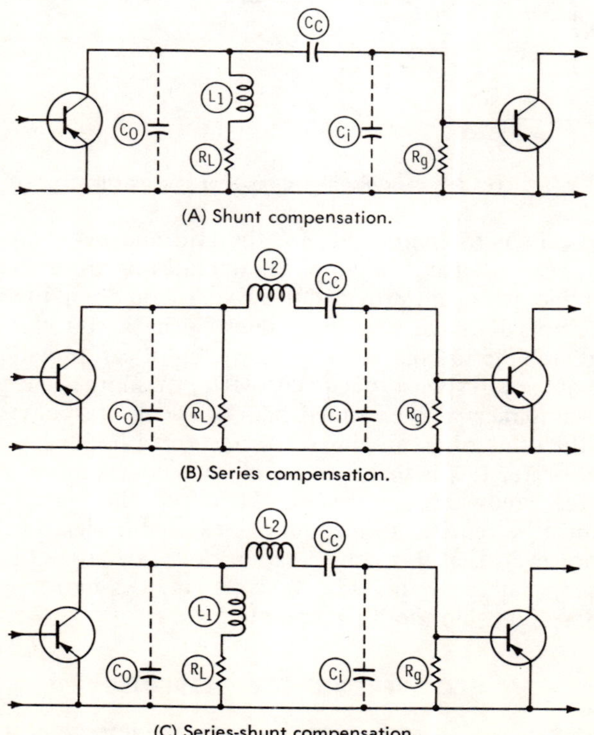

(A) Shunt compensation.

(B) Series compensation.

(C) Series-shunt compensation.

Fig. 5-4. Basic peaking-coil arrangements in a transistor amplifier.

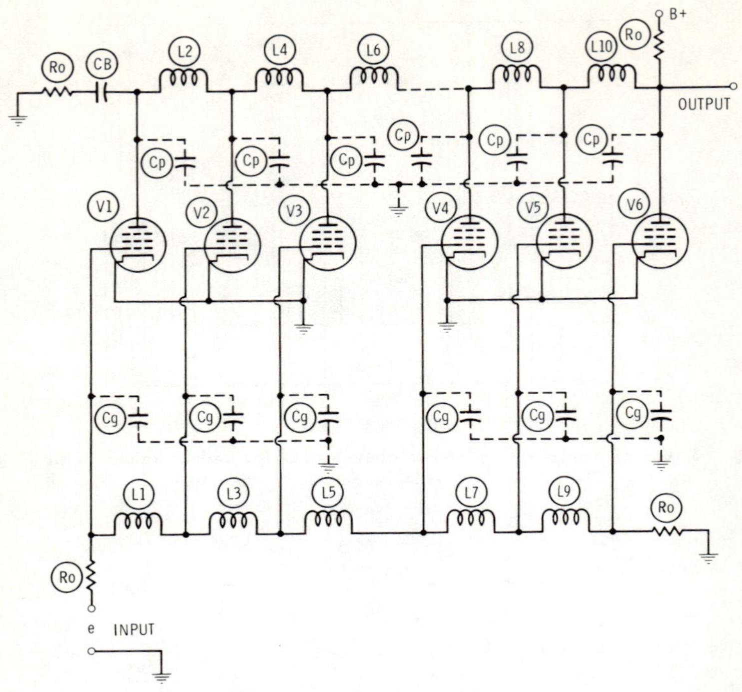

Fig. 5-5. Basic distributed-amplifier configuration.

peaking coils is to compensate for the grid and plate capacitances of the tubes. In effect, the amplifier operates as an artificial transmission line, with gain provided at each section along the line. The load resistors R_o have very low values, such as 100 ohms. A distributed amplifier is complex and costly, but it can provide uniform high-frequency response out to 50 MHz, or more. The gain of a vertical amplifier with extended high-frequency response tends to be less than that of an amplifier with restricted frequency response. The reason for this is that random noise becomes more of a problem as the bandwidth is increased. Therefore, the sensitivity of the scope must be reduced to maintain a reasonable signal-noise ratio. Note, however, that this consideration does not apply to sampling oscilloscopes because this type of scope employs narrow-band amplification following the input sampler.

LOW-FREQUENCY RESPONSE

Good response to low frequencies is equally important if accurate representation of waveforms is to be obtained. The low-frequency response of an amplifier depends largely on the size of the filter, bypass, and coupling capacitors. Generally, the larger the value of these capacitors, the better the response to low frequencies. The response can be extended to zero frequency (direct current) by using direct-coupled amplifiers like the ones shown in Fig. 5-6.

The usual practice is to design the vertical amplifier to have as wide a frequency range and as great a sensitivity as economically feasible; however, most of the time the horizontal amplifier will be used to amplify the sawtooth signal from the sweep generator of the oscilloscope. Normally, this signal is larger than the signal applied to the vertical amplifier, and its frequency range is usually less; therefore, the horizontal amplifier of an oscilloscope will commonly be designed to have less sensitivity and a narrower frequency response than the vertical amplifier, although in some models they may be identical.

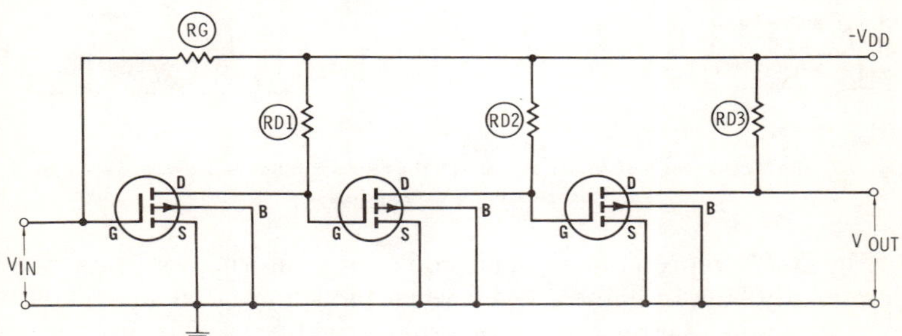

Fig. 5-6. A single-ended dc amplifier using p-channel enhancement-type MOS transistors.

DC AMPLIFIERS

An increasing number of oscilloscope designs include dc amplification for one or both deflection amplifiers. If only one amplifier is dc coupled, it will usually be the vertical since this choice will benefit the user more. Among the advantages offered by dc amplification are minimum phase shift and the extension of the low-frequency response to zero frequency (direct current). A few examples will be given of uses of dc oscilloscopes in servicing tv receivers.

Generally, the dc amplifier will be useful when viewing an ac waveform superimposed upon a dc component. The dc voltages alone can also be indicated, but these are usually more conveniently measured with the conventional voltmeter. Television circuits where the technican might find the dc amplifier particularly useful for spotting improper operation are the agc returns, direct-coupled video amplifiers, sync circuits, and dc restorers. As an example, assume that a service technician wishes to check the operation of a dc restorer. The dc amplifier input of the scope is connected to the restorer output, and the video signal being applied to the picture tube is observed. This signal might appear as in Fig. 5-7A. The scope is synchronized to show two horizontal sync pulses with video information appearing between them. When such a varying signal is passed through an amplifier that is ac coupled only, the signal tends to align itself about the ac axis. Furthermore, equal areas of the waveform will appear above and below the ac axis.

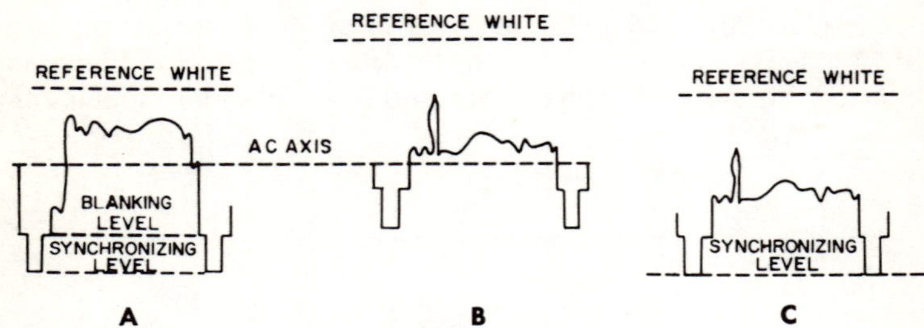

Fig. 5-7. The appearance of video signals at the picture tube as viewed on an oscilloscope having dc amplifiers.

Fig. 5-7A represents a signal containing information of a predominantly high brightness level; when the transmitted signal shifts to representing darker objects or lower brightness levels, it appears as in Figs. 5-7B and 5-7C. Fig. 5-7B shows that with an ac-coupled amplifier the signal averages itself about the ac axis, and the picture information assumes a position representing a brightness higher than the true level of the scene. A properly functioning dc restorer would restore the sync tips to the level shown in Fig. 5-7A, and along with them the picture information would assume its true level (Fig. 5-7C). Therefore, to check for proper operation of the dc restorer, it is only necessary to connect the dc amplifier of the scope to the modulated element of the picture tube and observe the level of the sync tips as the transmitted scene changes from light to dark. For proper operation the sync level should remain unchanged. A large percentage of present-day monochrome tv receivers do not employ the dc restorer, but the technician will find it in some of the earlier color-tv receivers; consequently, the aforementioned application of the dc amplifier in an oscilloscope may prove useful.

Another point of application for the dc amplifier might be at the video detector of a receiver. A great many receivers have the detector coupled directly to the grid of the first video-amplifier stage, and the instantaneous grid voltage produced by the video signal with respect to the cathode of the video amplifier is important when troubleshooting such stages for overloading or improper operation. A dc scope applied to these stages will aid in localizing such troubles.

For proper operation, sync-separated circuits are also critical with respect to the magnitude of signal. Too large a signal can easily result in unwanted video information being passed along with the desired sync signals. The dc scope will again prove useful in determining the instantaneous voltages of a signal at various points.

Fig. 5-8. An oscilloscope that has dc-coupled vertical and horizontal amplifiers.

Courtesy Lectrotech

The excellent low-frequency response of the dc amplifier is valuable in audio applications such as checking audio-amplifier response or the quality of the square-wave signal so often used in audio testing. Lack of phase shift at low frequencies allows the audio technician to determine that any square-wave tilt observed is due to the audio amplifier rather than to the square-wave source or scope amplifiers. Fig. 5-8 illustrates an oscilloscope that has dc vertical and horizontal amplifiers.

PUSH-PULL AMPLIFIERS

As was mentioned, most present-day oscilloscopes incorporate push-pull amplifiers. If the amplifiers do not have push-pull operation for all stages, at least the stages driving the deflection plates will have such operation. The advantages of this type of operation in oscilloscopes are similar to those obtained from push-pull operation in other applications: better hum cancellation, reduction of second harmonic distortion, and greater signal drive for the same supply voltage. In addition, the number of defocusing and trapezoidal effects are reduced because neither plate of each pair of deflection plates is at ground potential, as one is with single-ended operation.

ATTENUATORS

Since voltage amplifications as high as 1000 times are applied to the weakest signal being viewed, some means must be provided to prevent overloading when stronger signals are being viewed. In most cases, two attenuators will be used for this purpose—one a continuously variable control and the other a switching arrangement attenuating the signal in units of ten. The switch is placed at the amplifier input, and the variable control is usually separated from it by at least one amplifier stage. The operator should keep in mind that, with this arrangement, it is possible to overload the first amplifier stage and thus get a distorted response curve even though the vernier control is set for minimum response; consequently, a stronger signal should be reduced by the attenuator at the amplifier input, and minor adjustments should then be made with the vernier control.

The step attenuator for the vertical amplifier of one oscilloscope is shown schematically in Fig. 5-9A. In Fig. 5-9B the switch has been eliminated and only the active components for a single switch position are shown. R1 and R4 in series form a voltage divider, and only about 1/1000 of the vertical-input signal is applied to the grid of V1A. C4 is a trimmer providing frequency compensa-

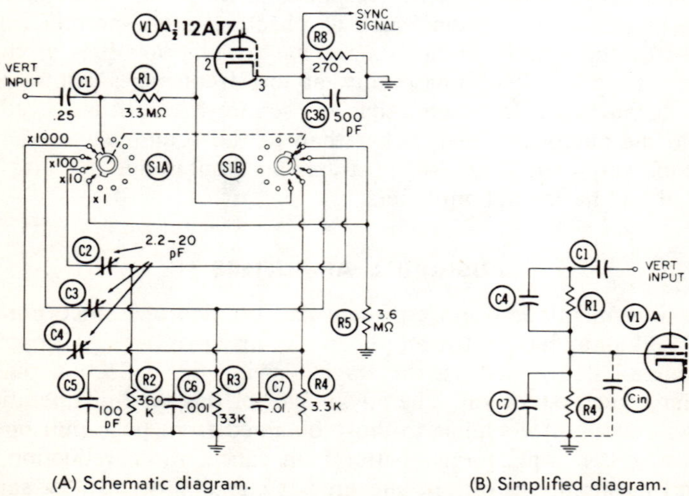

(A) Schematic diagram.

(B) Simplified diagram.

Fig. 5-9. A frequency-compensated attenuator.

tion for the applied signals. If a simple divider consisting only of R1 and R4 were used, the input capacitance (C_{in}) of V1A would bypass the higher frequencies more than it would the lower frequencies; consequently, the signal amplification applied to the vertical input would be reduced at the higher frequencies. To compensate for this effect, C4 is added. When the product of R1 × C4 equals the product of R4 × C_{in}, the voltage division is independent of frequency.

Note that the values of R2, R3, and R4 in Fig. 5-9A have been chosen to give an attenuation of 10 to 1, 100 to 1, and 1000 to 1 when used with R1. Note also that as R2, R3, and R4 decrease by a factor of 10, C5, C6, and C7 increase by the same factor to maintain a constant RC product. Theoretically, it should be possible to design an attenuator circuit without the use of C5, C6, and C7, with C_{in} serving in their stead. Then, since C_{in} does not change, there would be a different time constant for each attenuator position, R2 C_{in} being 10 times R3 C_{in}, and 100 times R4 C_{in}. To match these time constants, C2 would be adjusted to 1/10 C_{in}, C3 to 1/100 C_{in}, and C4 to 1/1000 C_{in}. Now, if we assume a value of 10 pF for C_{in} (2.2 pF for a 12AT7 tube plus stray capacities) the values of the adjusted trimmers would be 1 pF, .1 pF, and .01 pF, respectively. These values are perfectly all right in theory, but not very practical. When C5, C6, and C7 are added, the adjusted trimmers for the circuit of Fig. 5-9A should have a calculated value of about 10 pF each, a value within practical limits.

INPUT IMPEDANCES

We have mentioned that the oscilloscope has a relatively high input impedance and that this property is desirable in a voltage-measuring instrument. Fig. 5-10 shows how more accurate results are obtained if high-impedance equipment is used to measure high-impedance circuits. An open-circuit voltage measurement is one of the simplest to make and is pictured in Fig. 5-10A. The voltage source is represented by a voltage, E, in series with the internal

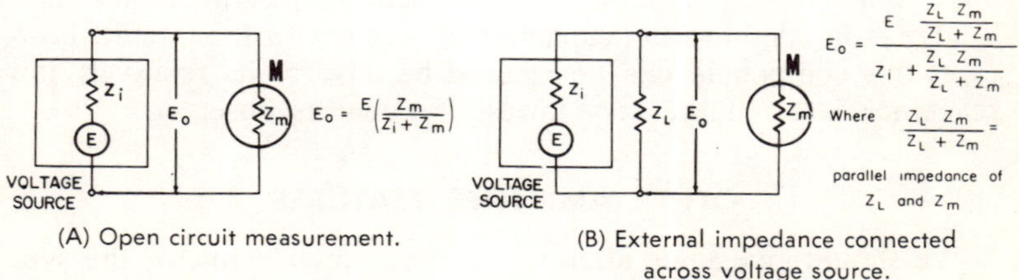

(A) Open circuit measurement.

(B) External impedance connected across voltage source.

Fig. 5-10. Diagram illustrating importance of using high-impedance equipment when measuring across high-impedance circuits.

impedance, Z_i. The measuring device, M, can be either a voltmeter or an oscilloscope. Now, a peculiar situation arises; once M has been connected to the voltage source, there is no longer an open circuit. How then can the open-circuit voltage be measured in this manner? The answer is that as impedance Z_m of the measuring device approaches infinity, the closed circuit comes closer and closer to open circuit conditions, and the measured voltage approaches the exact value of the source voltage (E). The voltage, E_o, across Z_m can be found by the following formula:

$$E_o = E\left(\frac{Z_m}{Z_i + Z_m}\right)$$

This formula shows that as Z_m becomes greater and greater, E_o approaches E in value. Of course, if Z_i is a very small value, Z_m does not have to be very large in absolute value to be relatively large compared to Z_i. In other words, when low-impedance circuits are to be measured, a very high impedance instrument is not necessary, although it can do no harm to use one if available. If conditions are reversed and Z_i is large compared to Z_m, more voltage will be dropped across Z_i, and the voltage reading E_o will be lower than the actual open-circuit voltage. In most cases there will be some external impedance connected to the voltage source, as in Fig. 5-10B, and the voltage across this impedance is the one to be measured. Z_m should be large compared to Z_L to avoid any change in voltage across Z_L as M is connected. The two impedances (Z_L and Z_m) will be in parallel, and their total value will be less than the smaller of the two. More voltage will be dropped across Z_i, causing an error in the voltage measured across Z_L.

The ideal input circuit for an oscilloscope would exhibit infinite resistance and no capacitance across the input terminals. In actual practice, of course, this condition can only be approached. Most general-purpose scopes have impedance ratings ranging from 1 to 5 megohms, with 25- to 50-pF capacitance across the input. These ratings show that the impedance term usually found in a manufacturer's specifications will be expressed in series resistance and parallel capacitance. The ac input of an oscilloscope will usually have a large value dc-blocking capacitor in series with the circuit; however, this component can be ignored because of its relatively low reactance, even at the very lowest frequencies observed.

OTHER AMPLIFIER FEATURES

We should give some attention to the takeoff point for the sync signal. If this signal is to be obtained from within the vertical amplifier circuit, several points could be used. The one shown in Fig.

5-9A will give a minimum amount of loading on the vertical amplifiers because it is a point of low impedance. This point is also ahead of the vernier control, and adjustment of the vernier control to obtain a larger or smaller response will not upset the synchronization. Since the takeoff point is at the front end of the amplifier, no amplification is obtained. Some oscilloscopes might require amplification of the sync signal before the signal is applied to the sweep generator. When a choice of polarity of the sync signal is offered, the desired polarity is sometimes obtained by means of a double-throw switch. A potentiometer of a suitable value is often bridged across two points having opposite polarities, and the position of the slider will control both the polarity and the amplitude of the sync signal.

Some oscilloscopes have provision for an inversion of the response curve on the screen. This is useful to those operators who like to compare a response curve directly with that pictured in a manual or service chart. Such inversion can be easily accomplished with push-pull amplifiers if double-pole, double-throw switches are used in the output to the vertical deflection plates. If the operator is interested in whether a positive signal will give an upward or a downward deflection of the trace, he can check this by touching the input cable first to a ground point and then to a point of known polarity. If the oscilloscope has dc amplifiers, the trace will move in one direction and will stay there. If an ac amplifier is being used, the trace will jump momentarily in the direction of deflection for that polarity and then return to its original position.

EXPANDED SWEEPS

The gain of the horizontal amplifier and the amplitude of the sawtooth signal in most modern oscilloscopes are such that the horizontal sweep can be expanded considerably. By merely adjusting the horizontal gain control, the operator can obtain a sweep 10 times or more the width of the oscilloscope screen. In this manner, small portions of a response curve can be examined in detail.

Some oscilloscopes incorporate a special circuit for obtaining a greater expansion than the horizontal amplifier affords in normal operation. When such models are operated in the expanded sweep position, a greater portion of the sawtooth signal developed by the sweep oscillator is fed to the horizontal amplifier. The horizontal amplifier is overdriven by this signal, but the middle portion of the sweep is linear and is greatly expanded. The linear portion of the sweep can be shifted by means of a control that varies the bias on the horizontal amplifier stage to which the sawtooth signal is applied.

RISE TIME—WRITING SPEED

Manufacturers use a number of terms to describe the performance of their oscilloscopes; some of these are well-known because of common usage, while others are lesser known and might need some explanation. Rise time is a term that has been used sparingly in the past but is seen more frequently now. It refers to the time it takes the oscilloscope amplifiers to respond to a signal that changes instantaneously from zero to its final value. At one instant no signal is applied to the amplifier, and, of course, no output is obtained. At the next instant the full amplitude of signal is applied, but it takes a definite interval of time for the amplifier output to rise to a final value. This interval, minus 10% at the beginning and 10% at the end, is the rise time of the amplifier. A diagram illustrating rise time is shown in Fig. 5-11. Rise time is usually quoted in microseconds, and as a figure of merit the smaller values of rise time will generally indicate better performance in an oscilloscope. Short rise time shows up in the ability of the scope to reproduce faithfully a waveform having steep leading edges. This requires a good response to high frequencies.

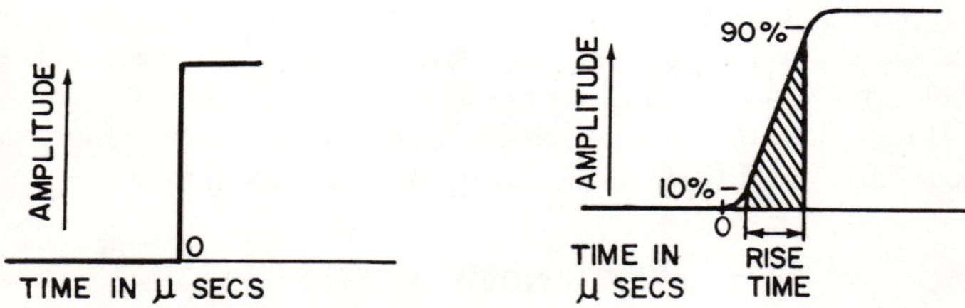

Fig. 5-11. Diagram illustrating rise time.

One manufacturer gives the following formula for calculating the high-frequency response necessary to reproduce satisfactorily a wavefront of given rise time.

$$f = \frac{.36}{t}$$

where,

f is the frequency in megahertz,
t is the rise time in microseconds.

Suppose the rise time of an oscilloscope vertical amplifier is one microsecond. Substituting this value for t in the formula gives a high-frequency response value of 360 kilohertz.

Writing speed can be subdivided into three classifications: beam writing speed, photographic writing speed, and sweep writing speed.

Beam writing speed refers to the linear speed of the beam as it travels across the face of the cathode-ray tube, regardless of direction. Thus, the beam writing speed would be greater with a vertical signal applied than without, because, in addition to horizontal movement due to sweep-circuit voltage, vertical movement has been added by the signal, but with no increase in time for a sweep cycle. For the same reason, beam writing speed would be increased if the sweep amplitude were increased by means of the horizontal gain control. Any increase in beam speed is usually accompanied by a decrease in the trace intensity unless some attempt to counteract this effect is made.

Photographic writing speed is of interest mainly to determine the limits for making photographs of waveforms. These limits depend on a number of factors such as film emulsion speed, lens aperture, tube phosphor, and accelerating voltage. The term photographic writing speed is seldom encountered in service literature and will not be given further mention here.

Sweep writing speed refers to the speed of the beam in a horizontal direction due to sweep voltage applied to the horizontal amplifier. This speed increases with increased sweep frequency and horizontal gain. Therefore maximum sweep writing speed will be obtained when the sweep frequency and the horizontal gain controls are at their highest setting.

Writing speeds are usually stated in inches per microsecond or sometimes in reverse order, microseconds per inch. For example,

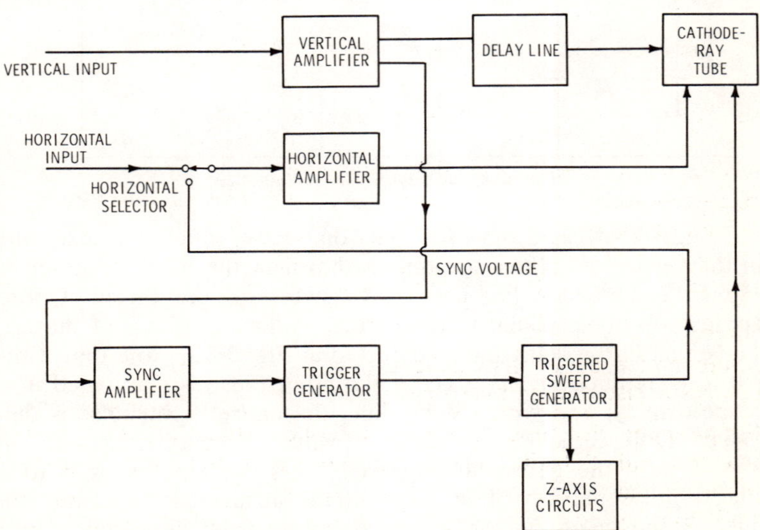

Fig. 5-12. Lab scopes include a delay line in the vertical channel.

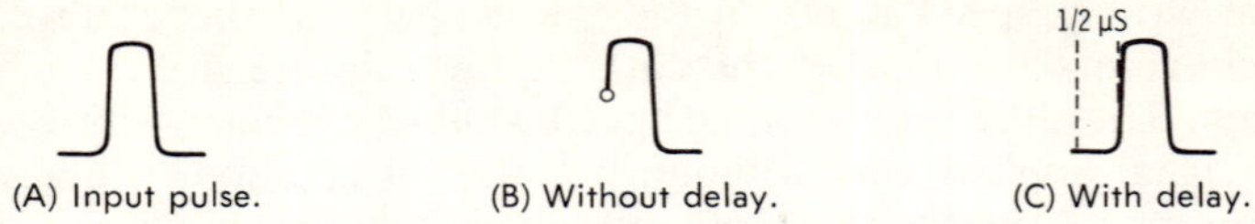

Fig. 5-13. Function of a delay line in the display of a fast pulse.

one manufacturer states that his oscilloscope has a maximum usable writing rate of 0.03 second per inch to 0.3 microsecond per inch, and another quotes a maximum speed of 1 inch per microsecond for the triggered sweep of his oscilloscope. This information can be useful to evaluate possible scope performance in the following manner. Suppose you consider buying an oscilloscope with a maximum writing speed of 4 inches per microsecond. This tells you that if you can synchronize a 1-MHz signal at the maximum sweep rate, one cycle of this signal will be spread across 4 inches of screen. At the same rate, one cycle of the 3.58-MHz color burst would cover a little more than 1 inch of screen; a horizontal sync pulse of 5-microsecond duration would be expanded to an equivalent of 20 inches of screen, and so on. In many cases it is desirable to expand a response curve as much as possible in order that fine detail on the curve can be seen. This is especially true when the operator must view a large number of cycles of a particular signal having a frequency much higher than that of the sweep.

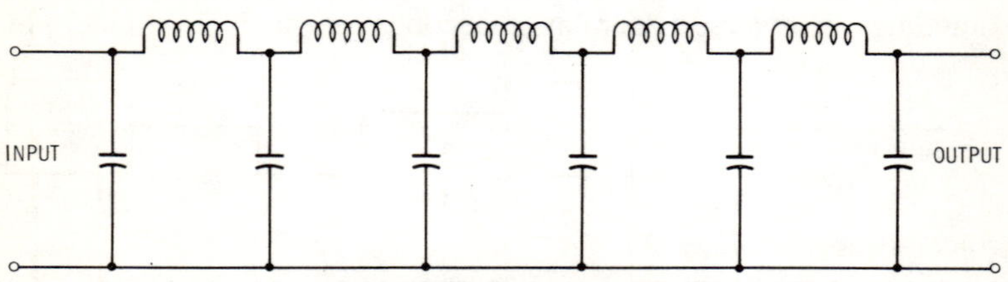

Fig. 5-14. Basic delay-line configuration.

Triggered-sweep scopes with fast time bases employ a delay line in the vertical-amplifier section, as shown in the block diagram of Fig. 5-12. The delay line permits the sweep to "get started" before application of the signal to the vertical-deflection plates of the crt. The effect of the delay line is depicted in Fig. 5-13. Note that without a delay line, the first part of the input pulse is eliminated in a scope display, and with a delay line, the entire input pulse is displayed. Different scopes use different values of delay, depending on how rapidly the particular triggered-sweep system responds to a trigger pulse. A delay line is essentially an artificial transmission line, as shown in Fig. 5-14. An elaborate delay line may utilize several dozen sections.

Chapter 6

Special Features

At various points throughout this book, we use the concept of the basic general-purpose oscilloscope. Such an oscilloscope would include the cathode-ray tube; high- and low-voltage supplies; intensity, focus, and positioning controls; a horizontal sweep system with frequency controls; provision for synchronization; horizontal and vertical amplifiers; and some provision for controlling the amplitude of the horizontal and vertical signals.

An oscilloscope limited to these features is seldom seen. Usually, a number of other features are added to increase the usefulness and efficiency of the instrument. Some of these have already been discussed but are included in the following list, along with others that may not have been mentioned before. Some oscilloscopes may contain a majority of the features listed, although it is probable that no single oscilloscope will contain them all.

1. Expanded sweep.
2. Driven sweep.
3. Slow-speed sweep.
4. Line-frequency sweep.
5. Fixed-frequency sweeps for viewing signals with 60-hertz and 15,750-hertz rates.
6. Sync at line frequency and at two times line frequency.
7. Input for external sync signals.
8. Automatic sync-level control.
9. Voltage-calibration circuit.
10. Rf detector circuit.
11. Polarity reversal of vertical deflection.
12. Positive or negative sync signal.
13. Intensity modulation.
14. Sawtooth output signal.

15. Retrace blanking.
16. Phasing control of line-frequency sweeps.

Expanded sweeps and driven sweeps have been discussed in preceeding chapters. Slow-speed is provided on some oscilloscopes by a front-panel jack connected to the frequency-determining network inside the scope. To use this feature, the operator sets the sweep frequency switch to the proper position and connects a large value of capacitance from the front-panel jack to ground, thus lowering the frequency of the sweep oscillator. A good quality capacitor should be used because any leakage would tend to raise the sweep frequency. This is contrary to the effect of the capacitance, and if the leakage is sufficiently great, the net result may be a sweep rate actually higher than before the capacitor was added.

FIXED AND LINE-FREQUENCY SWEEPS

Sweeps at line frequency and at the horizontal and vertical rates of a tv receiver are commonly used by the technician. The line-frequency sweep is obtained by feeding a signal to the horizontal amplifier, taken from some winding on the power transformer. Therefore, it is not a sawtooth but a sine-wave sweep. It can be used with an rf sweep generator to develop a response curve of a receiver, provided the oscilloscope also has a phasing control (item 16 in preceding list). If this latter feature is omitted from the oscilloscope, the operator should use the synchronized sweep signal from the rf generator to develop the oscilloscope sweep or else be prepared to accept a double response curve and some probable waveform distortion.

Item 5 refers to features obtained by special positions of the sweep frequency switch. These positions are usually marked V and H TV, or in some similar manner, and sweep rates of 30 and 7875 hertz, respectively, are produced when these positions are used. A display of two cycles of signal at the tv vertical or horizontal sweep rate is obtained in this manner, with minimum adjustment of the sweep frequency controls.

SYNCHRONIZATION REFINEMENTS

Items 6, 7, and 8 are useful in obtaining stable synchronization under trying circumstances. As an example of how these features can be used, suppose that the technician wishes to view the video signal in a tv receiver as it occurs between two vertical sync pulses, but some fault in the receiver has attenuated the vertical sync pulses at the point where the oscilloscope is connected. If the sync control of the oscilloscope is set to INT (internal) position, synchronization may be difficult to obtain because the sync signal is taken from the signal being viewed, and the necessary vertical sync pulses have been attenuated or are missing from the signal at this point in the receiver circuit. If the sync control is set at the LINE position, sync signals of the proper amplitude and frequency are automatically provided from a point within the oscilloscope.

If trouble is encountered with a signal that does not recur at line frequency or some multiple thereof, the sync control can be set to EXT (external) position and the sync signal can be taken from some point in the receiver circuit where a more definite pulse is obtained.

With regard to item 8, it was mentioned earlier that either too strong or too weak a sync signal results in poor synchronization; therefore a feature automatically providing the correct level of sync signal will do much toward simplifying the synchronization problem. A limiting stage somewhere in the sync circuit will give a constant-level sync signal, although the vertical signal level may vary considerably at the same time. This feature is provided in some oscilloscopes by the manufacturer.

VOLTAGE CALIBRATION

Voltage-calibration circuits (item 9) enable the operator to measure the amplitude of the signals shown on the oscilloscope screen. This is one of the more important features the instrument can have, since it is primarily a voltage-measuring device. Not only the overall or peak-to-peak voltage of a waveform can be measured, but any detail on the waveform can be compared to other details or to the maximum voltage. This is done by first noting the amount of deflection obtained for a known voltage and comparing it with the deflection obtained for an unknown voltage. One of the simplest provisions for measuring voltages with a scope is to connect a jack on the front panel to some point of known voltage inside the scope case. A 6.3-volt ac filament winding will provide a sine-wave signal of approximately 18 volts peak to peak. If this voltage is connected to the front panel jack, it is then available for connection to the vertical input for comparison to some voltage to be measured. This is the type of calibration provision shown in Fig. 6-1. Fig. 6-2 shows how it can be used to measure the peak-to-peak amplitude of a composite video signal. The 1-volt calibration signal occupies a vertical span of six horizontal lines, whereas the composite video signal occupies two lines or ⅓ of 1 volt. Therefore, the peak-to-peak voltage of the video signal is ⅓ volt. If necessary, finer divisions of the calibration grid can be used for more accurate results. Rather than have the calibration voltage terminate on the

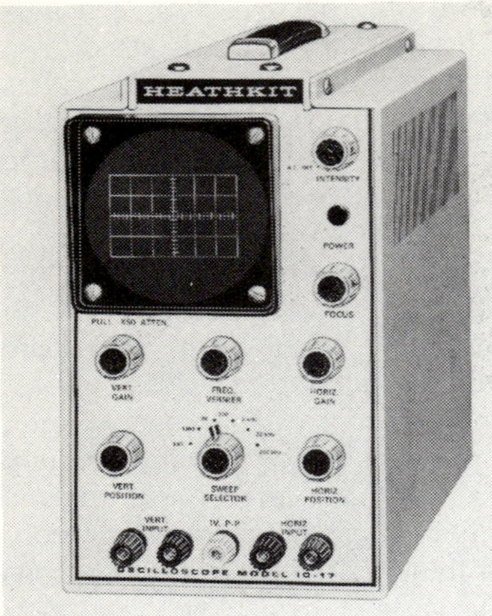

Courtesy Heath Co.

Fig. 6-1. 1-volt peak-to-peak calibrating voltage source located on front panel of oscilloscope.

front panel, some scopes use a switching arrangement to apply the test signal to the vertical system. The switch controlling the calibration voltage will often be found as part of another adjustment, such as the sync selector switch pictured in Fig. 6-3. With the switch of Fig. 6-3 in the calibrate position, a 10-volt, peak-to-peak signal is automatically applied to the vertical attenuator input circuit.

When a number of calibrating voltages are available, they are usually allotted a separate panel control, as in Fig. 6-4, where a choice of four voltages in decade steps is provided. An example of a somewhat more elaborate method of internal scope calibration is where the instrument incorporates a peak-to-peak reading voltmeter. A front panel view of such an instrument is shown in Fig. 6-5. The vertical attenuator in this example has two calibration positions (3 and 10 volts) corresponding to two scales on the meter. With the vertical attenuator in either of these two positions,

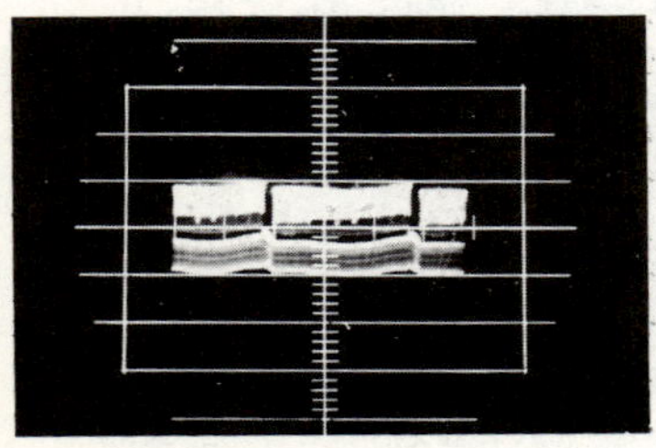

(A) Composite video signal.

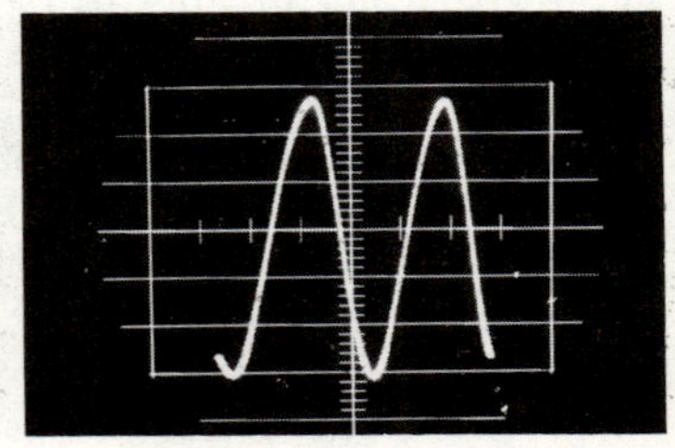

(B) 1-volt calibration pattern.

Fig. 6-2. Photographs taken directly from the screen of a typical service scope.

Fig. 6-3. Sync selector includes CAL position for internal calibrating voltage.

the calibrating voltage knob selects any voltage from 0 to 3, or 0 to 10 volts, and the selected voltage is indicated by the meter pointer. At the same time, a sinusoidal waveform is shown on the screen with an amplitude corresponding to the selected voltage. The final amplitude of the waveform also depends on the setting of the vertical gain control. When the calibration waveform is compared with a waveform to be measured, the position of the vertical attenuator must be considered. Thus, if a 5-volt calibrating waveform is adjusted with the vertical gain control for a vertical deflection of two inches and the deflection from an unknown voltage is also two inches when the vertical attenuator is set at position 10, the un-

Fig. 6-4. This oscilloscope permits the use of four different calibrating voltages.

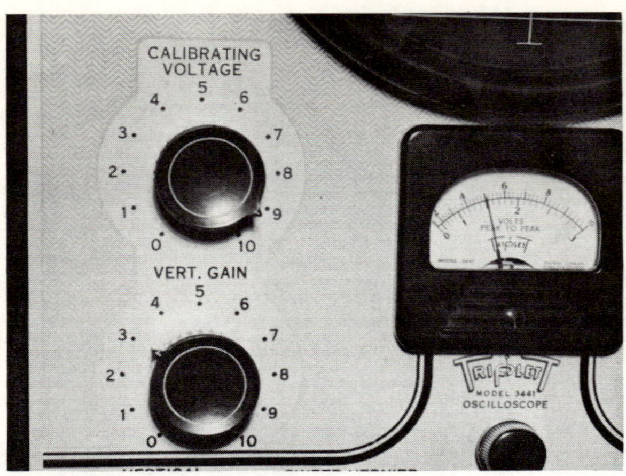

Fig. 6-5. Internal calibration features which include peak-to-peak reading voltmeter.

known voltage would be 10 × 5, or 50 volts. Thus, it can be seen that this particular instrument offers a calibration range from 0 to 1000 peak-to-peak volts. The technician should keep in mind then that any accessory probe used with a scope may introduce a certain amount of signal attenuation, and this must be considered if accurate peak-to-peak measurements are to be realized.

Triggered-sweep oscilloscopes generally provide calibrated vertical step attenuators as pictured in Fig. 6-6. In other words, the step attenuator indicates p-p volts per vertical division of the grati-

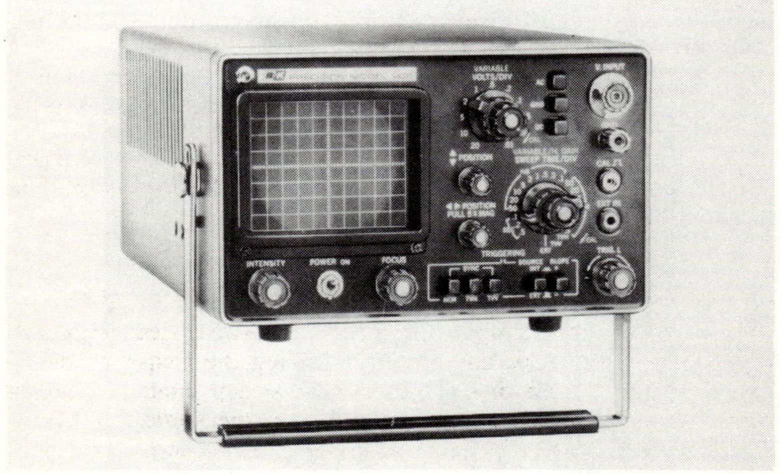

Courtesy Dynascan Corp.

Fig. 6-6. Vertical step attenuator indicates peak-to-peak volts/div.

cule. In the present example, a range from 0.01 to 20 volts/div is provided. As noted previously, if a 10-to-1 probe were in use, this range would become 0.1 to 200 volts/div. Note that a calibrated step attenuator indicates correctly when the variable vertical-gain control is set to its reference position. On the other hand, if the variable control is turned from its reference position, the step attenuator will indicate incorrectly. Note that a square-wave source of calibrating voltage is also provided on the front panel of the instrument. This source is useful to check the calibration of the vertical step attenuator in case it is suspected that the overall gain of the vertical amplifier might need to be adjusted.

RF DETECTORS

The oscilloscope is limited in direct viewing to signals within the frequency range of its amplifiers, but modulated rf signals can also be viewed with the oscilloscope if these signals are detected before being applied to the oscilloscope amplifiers. This detection can be accomplished by circuits within the oscilloscope (item 10 in preceding list) or by using the proper external probes. The latter method is more common. Since polarity reversal of vertical deflection (Item 11) and positive or negative sync signal (Item 12) have been discussed in previous chapters, no further mention will be made here.

INTENSITY MODULATION

Intensity (Z axis) modulation (item 13) can be used to determine the frequency of a signal or to measure the deviation of a trace or portion of a trace. An alternating signal of proper amplitude fed to the intensity-modulation jack will increase or decrease the trace intensity in step with the alternations and thus mark the trace at regular intervals. An alternating signal consisting of sharp pulses works best because it produces more distinct markings than an alternating signal of a smoother nature.

SAWTOOTH OUTPUT SIGNAL

Item 14 describes a signal taken from the sweep generating system of the oscilloscope and therefore having the frequency determined by the setting of the oscilloscope sweep controls. Such a signal can be used for signal substitution in the vertical and horizontal systems of a tv receiver. Although the sawtooth signal does not match the normal signal found in these systems, it can be used for troubleshooting purposes.

RETRACE BLANKING

Retrace blanking (item 15) is usually included as a feature of most oscilloscopes. In some examples the blanking operates continuously, and in others, it may be turned on or off by means of a switch. Usually the retrace period is an extremely small fraction of one sweep cycle and does not interest the oscilloscope operator. In such cases, retrace blanking eliminates any confusion or distraction that might result if the retrace were visible. When the retrace period occupies a greater portion of the cycle, it may be desirable to turn off the retrace blanking so that no part of the signal will be lost during retrace. The waveform visible during retrace may be more difficult to interpret than that shown on the normal portion of the sweep because the retrace is usually nonlinear in nature, but it is usually possible to determine facts of a quantitative nature, such as the number of cycles of signal lost during retrace.

PHASING CONTROL

The phasing control (item 16) is useful when a line sweep is used. At this position of the sweep controls, a sine-wave signal is taken from some point of the oscilloscope circuits (usually a winding on the power transformer) and used to drive the horizontal deflection system. The phasing control is used to vary the particular point on the sine-wave signal at which the sweep begins.

DUAL-BEAM DISPLAY

Electronic switches are sometimes used to obtain a dual-beam display. This permits two waveforms to be displayed one above the other on the scope screen. The electronic switch feeds the signal alternately into the vertical amplifier, and adds a different dc component to each signal. This is called an effective dual-beam function, because the crt operates practically as if it had two electron guns, and a pair of vertical-amplifier channels. Lab scopes may provide built-in electronic-switching circuits that require no external devices. A typical screen display is illustrated in Fig. 6-8.

A block diagram of a vertical amplifier with a built-in electronic switch is depicted in Fig. 6-7. Two preamplifiers are employed, which are switched on and off alternately. Outputs from the preamplifiers are fed into a conventional vertical-deflection amplifier. Note the delay line between the vertical-amplifier stages. A delay line is commonly used in triggered-sweep scopes to permit the horizontal multivibrator to get started before the vertical-input signal arrives at the crt. Thus, the delay line avoids loss of the

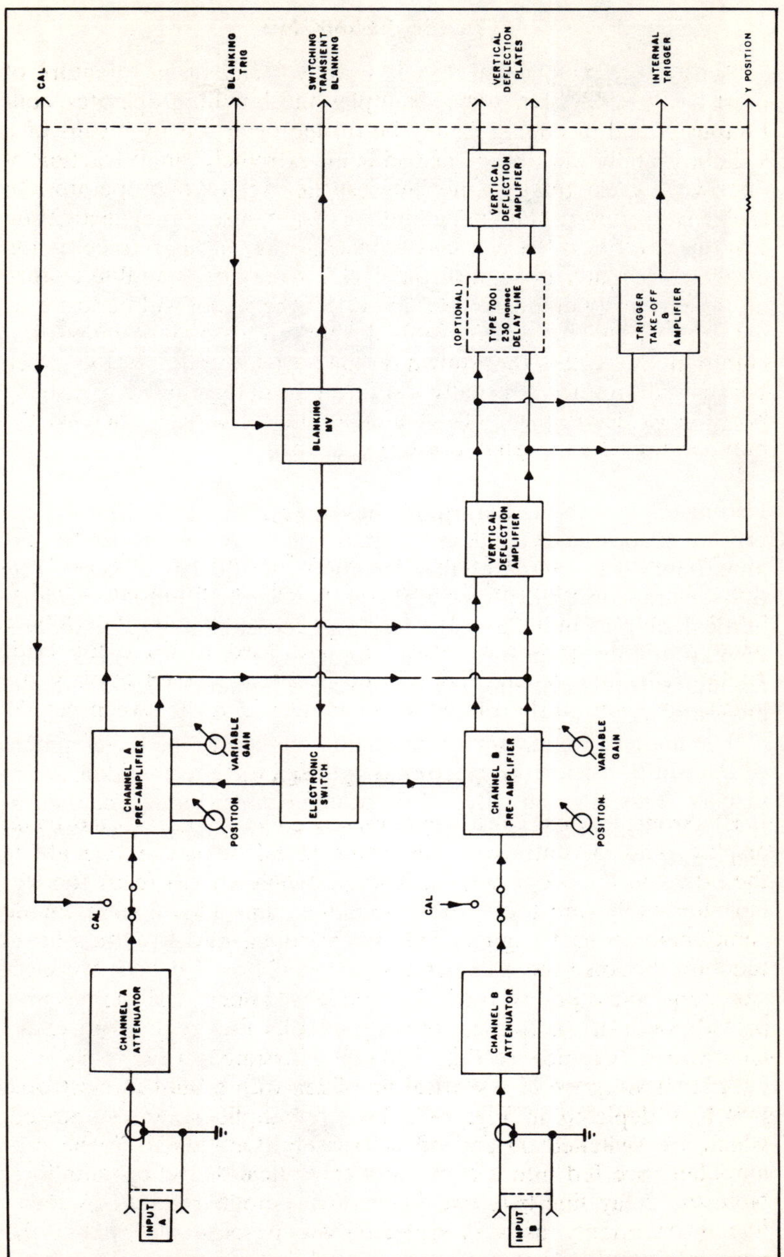

Fig. 6-7. A vertical amplifier with built-in electronic switch.

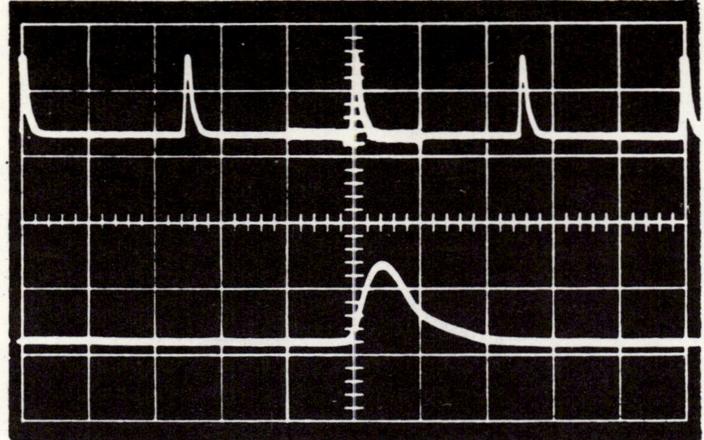

Fig. 6-8. A dual-beam display.

leading edge of the waveform in the displayed pattern. Many modern lab scopes have facilities for plugging in various types of preamplifiers. This increases the versatility of the basic scope. Fig. 6-9 illustrates a typical dual-trace plug-in vertical amplifier. Other vertical plug-in units provide for four vertical inputs and permit display of four independent waveforms, one above another. This facilitates troubleshooting of complex equipment, such as digital computers.

The modern trend to miniaturization has led to the development of built-in transistorized electronic switches for effective dual-beam display. This is essentially a free-running multivibrator configuration, with semiconductor diode clippers to provide a good square-wave output.

Fig. 6-9. A plug-in amplifier that provides an effective dual-beam function.

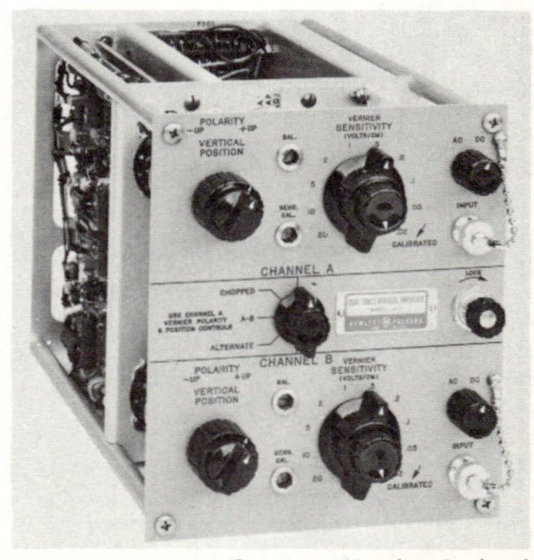

Courtesy Hewlett-Packard

Chapter 7

Accessories

The modern oscilloscope, with all its versatility, would seem to be complete in itself, needing no accessory equipment to increase its usefulness. Such, of course, is not true; a number of accessories are available and do extend the usefulness of any scope with which they are used. They range from small objects, like probes, to objects sometimes larger than the oscilloscope itself, like the electronic switch.

PROBES

The simplest type of probe (one that can hardly be called a probe, or an accessory either, for that matter) is the test lead. Test leads are simply convenient lengths of wire for connecting the oscilloscope input to the observation point. At the oscilloscope end they usually terminate with spade lugs, banana tips, or other tips to fit the input jacks of the scope, and at the other end there will be alligator clips or other convenient means for connecting to electronic circuitry. Since the oscilloscope input has high impedance and high sensitivity, the test leads should be shielded to avoid hum pickup, unless the scope is connected to a low-impedance, high-level circuit.

Although the input impedance of most scopes is considered relatively high compared to the circuits where they may be connected, it is often desirable to increase this impedance in order to avoid loading the circuits or causing unstable effects. The input capacitance of the scope, plus stray capacitance of the test leads, may be just enough to cause a sensitive circuit to break into oscillation when the scope is connected. This effect can often be prevented by an isolation probe made by placing a carbon resistor in series with the test lead, as shown in Fig. 7-1. Values of 10,000 to 47,000 ohms work well in many cases. A slight reduction in the amplitude of the

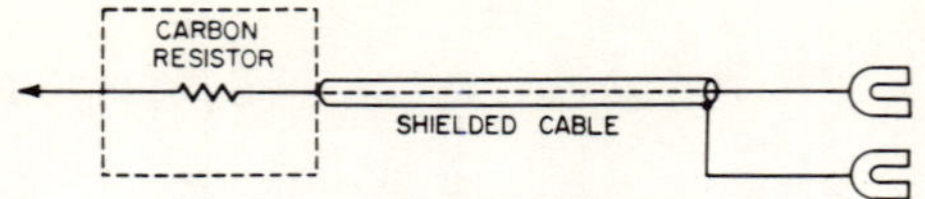

Fig. 7-1. A simple isolation probe.

waveform must be accepted with this system. The waveshape of a signal may also be changed by this probe. To avoid this possibility, a high-impedance, compensated probe can be used. This probe, often called a low-capacitance probe, is illustrated in Fig. 7-2. The probe itself is shown at Fig. 7-2A, and its circuit diagram at Fig. 7-2B. The theory of this probe is exactly the same as that for the compensated attenuator mentioned in Chapter 5. The probe will give minimum waveform distortion when adjusted so that the $R_1 \times C_t$ equals $R_2 \times C_{in}$, where, C_{in} is the total input capacitance of the oscilloscope. Typical values for R_1, R_2, and C_t are: R_1, 1 megohm; R_2, 110K; and C_t, 5 to 20 pF. These values will give about a 10-to-1 signal reduction.

The highest voltage that can be safely connected to an oscilloscope input is determined by the voltage rating of the input blocking capacitor. This rating is usually 600 volts or less. Some locations in tv receivers have voltages exceeding this figure. The entire high-voltage section from the plate of the horizontal output tube onward is an example. Waveforms in this section can be viewed through the use of a capacitive voltage-divider probe like that of

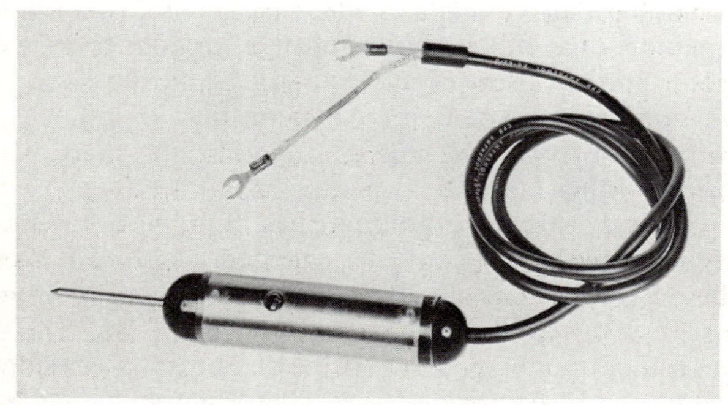

(A) Physical appearance.

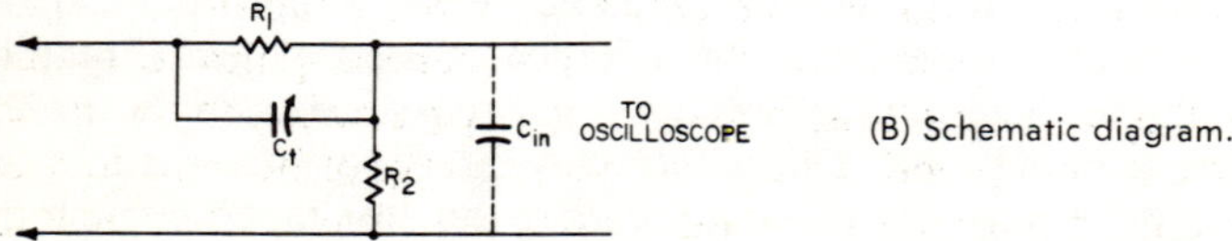

(B) Schematic diagram.

Courtesy Hickock Electrical Instrument Co.

Fig. 7-2. A high-impedance, frequency-compensated probe.

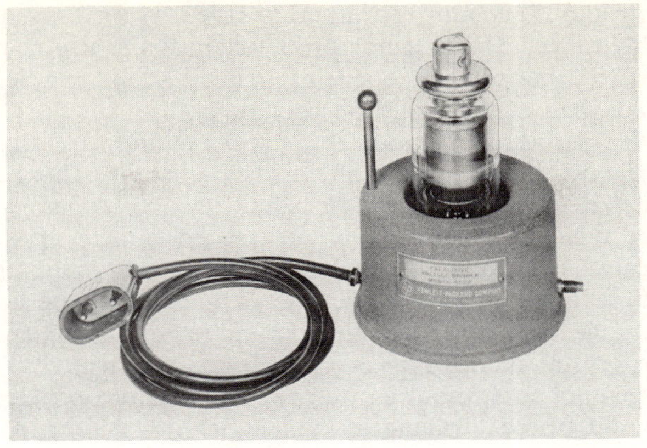

(A) Physical appearance.

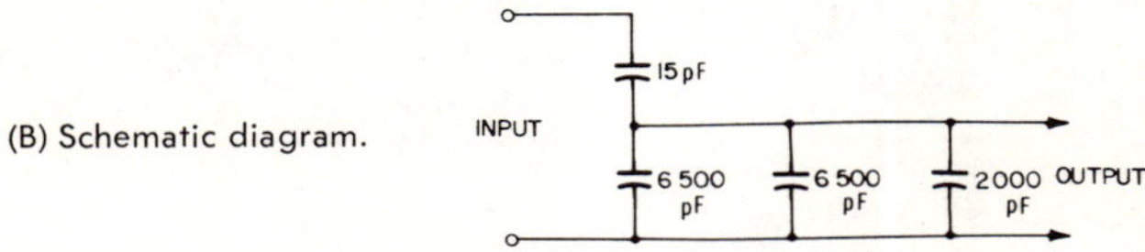

(B) Schematic diagram.

Fig. 7-3. The capacitive voltage divider for use with high-voltage waveforms.

Fig. 7-3A. This particular probe measures voltage up to 25 kilovolts rms. The probe illustrated is used primarily for laboratory applications, but indicates the wide variety of probes available. The schematic of Fig. 7-3B shows that the voltage divider consists of a 15-pF capacitor in series with three capacitors totaling 15,000 pF. This gives a voltage division of 1,000 to 1.

It is interesting to note the mechanical construction of this probe. The 15-pF capacitor is a vacuum capacitor with a very high voltage rating. One terminal of this capacitor furnishes the connection for the high voltage. A spark gap is provided to prevent damage to the divider from excessively high voltage. Such an elaborate probe is seldom required in the ordinary shop, but capacitive voltage-divider probes could be constructed for circuits containing much lower voltages. If such a project is contemplated, high-quality components should be used. Presence of resistance in the probe could cause integration or differentiation of the waveform.

The vertical amplifiers of the average general-purpose oscilloscope suffer a drop-off in response to frequencies above a few megahertz. Consequently, the oscilloscopes cannot be used to view directly any rf signal exceeding this upper limit. The standard broadcast frequencies, extending as they do from about 500 kHz to 1600 kHz, are well within the direct range of the scope amplifiers, but the i-f and rf sections of fm and tv receivers pass signals much above this range. The detector probe extends the useful range of the scope to cover these frequencies, and the scope can be used to signal trace these sections of the receivers.

Detector probes, or demodulator probes as they are also called, may be divided into two general classifications: (1) those employing crystal diodes and (2) those using vacuum tubes. The vacuum-tube type has been largely superseded by the crystal-diode type because the vacuum-tube type is larger and requires heater and plate voltages for operation. The vacuum-tube type does have the advantage of being able to amplify and to handle higher voltages than the crystal diode. However, high scope sensitivity and the small signal voltages found in the rf and i-f sections of the receiver tend to nullify these advantages.

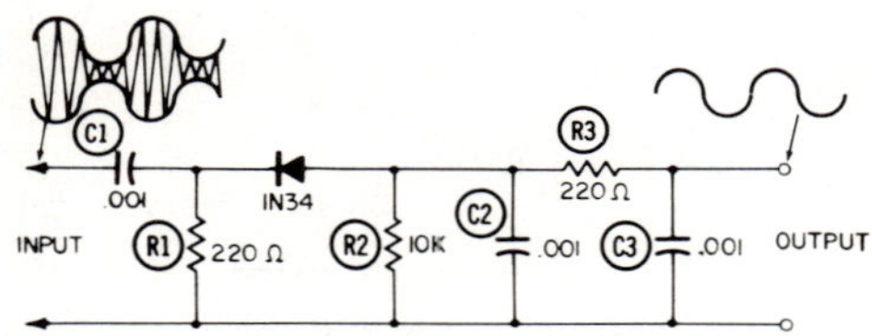

Fig. 7-4. A crystal-diode probe for demodulating rf signals.

Fig. 7-4 is the schematic for a demodulator probe using a crystal diode. C1 is a blocking capacitor that keeps any dc signal component from being applied to the crystal diode. R1 and R2 form a complete dc path for the crystal diode. R3, C2, and C3 together make a one-section filter to reduce the rf signal amplitude after it has been rectified by the diode. A modulated rf signal applied to the input of the demodulator probe reaches the output of the probe demodulated and filtered as shown in Fig. 7-4. Only the modulating signal remains. The probe can be simplified by omitting R3, C2, and C3, and it will still function satisfactorily at frequencies above the response limit of the scope.

A balanced demodulator probe is pictured in Fig. 7-5. This probe is useful when it is desired to measure the rf and also maintain a proper impedance termination at the point of application.

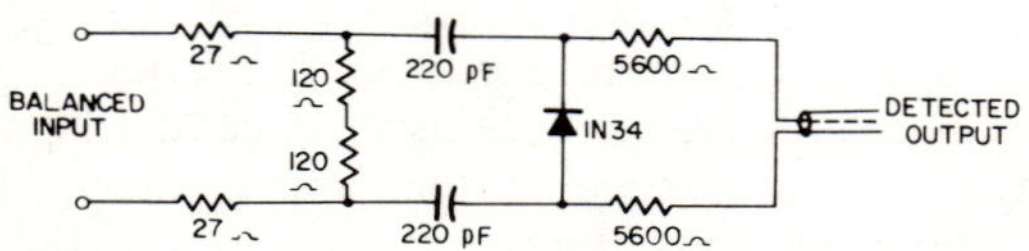

Fig. 7-5. A balanced input demodulator.

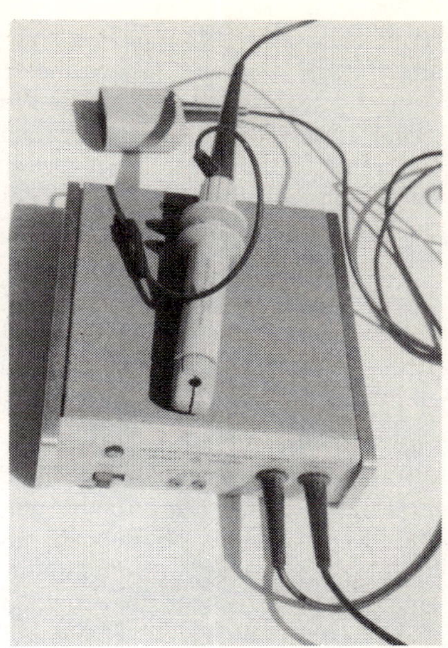

Fig. 7-6. A clamp-around current probe for displaying current waveforms.

The input circuit offers a balanced input of 300 ohms, while the output circuit offers an unbalanced one to match the oscilloscope.

Clamp-around current probes, such as pictured in Fig. 7-6, are used chiefly in laboratory work. Of course, a current probe can be used in electronic servicing, if desired. As an illustration, if the probe is clamped around a lead to the horizontal-deflection coils in a tv receiver, the current sawtooth wave-form will be displayed. This type of probe is called an active device. The probe itself contains a miniaturized current transformer, of which the lead being tested becomes the equivalent of a one-turn primary. Next, the output from the current transformer is stepped up by a transistor amplifier with self-contained batteries. The output from the amplifier is applied to the vertical-input terminals of a scope. A typical current probe has a sensitivity of 1 mV per mA, and a frequency range from 60 Hz to 4 MHz.

VOLTAGE CALIBRATORS

Built-in calibration circuits are mentioned in Chapter 6; they range from fairly simple to more complex examples. Several manufacturers supply separate self-contained units that can be used to calibrate any oscilloscope. A few of these units are pictured in Figs. 7-7, 7-8, and 7-9. All derive their operating power from the power line source. The unit in Fig. 7-7 is a Simpson Model 276 oscilloscope calibrator. It supplies a sine-wave signal to the oscilloscope and meters it at the same time. Every alternate position of the voltage selector switch is a "feedthrough" position, permitting any signal applied to the input terminals to feed directly through the calibrator to the scope input terminals. The other switch positions select voltage ranges from 1 to 250 volts peak-to-peak. Peak and rms values are also marked on the scale for each position. A calibrating voltage adjustment control allows the operator to vary the applied voltage from 0 to full scale for any voltage setting of the switch. Since this calibrator meters the output voltage directly, the accuracy of the unit is as good as the meter itself, regardless of line voltage variations.

The Hickok Model 630 television voltage calibrator is shown in Fig. 7-8. In this unit, the square-wave output from a multivibrator stage is attenuated to the desired level as indicated by the meter and then applied to the output terminals of the calibrator. Stability of level is obtained by using a regulated power supply. Four voltage ranges with maximum output values of .1, 1, 10, and 100 volts peak-to-peak are provided.

The Dumount Model 264-B oscilloscope calibrator (Fig. 7-9) has an interesting circuit design. A square-wave output is obtained through the use of clipped sine waves. This is accomplished in the

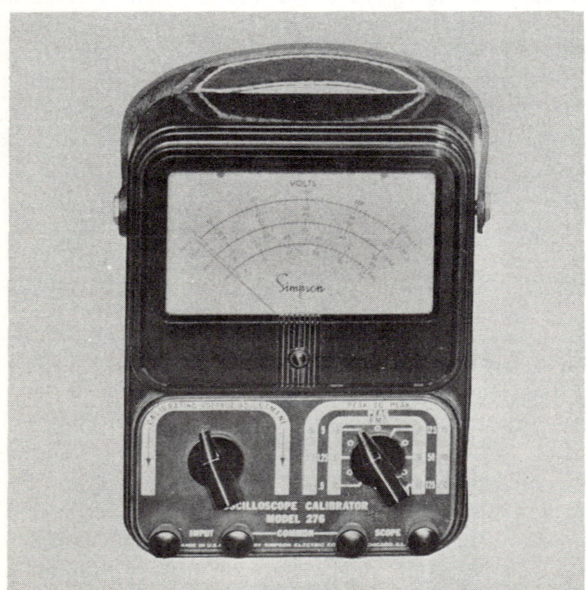

Courtesy Simpson Electric Co.

Fig. 7-7. Simpson Model 276 oscilloscope calibrator.

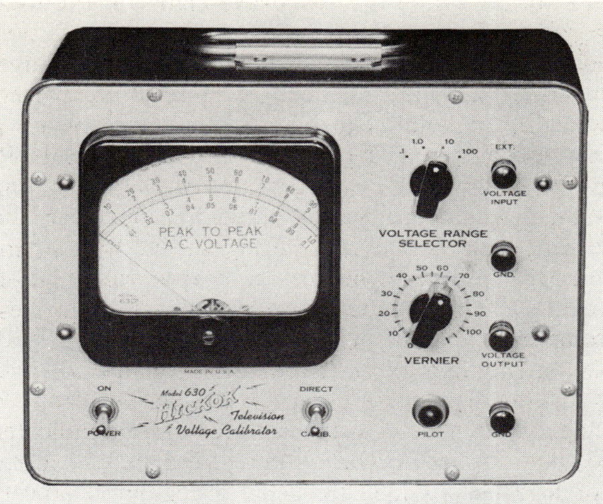

Courtesy Hickock Electrical Instrument Co.

Fig. 7-8. Hickock Model 630 television voltage calibrator.

following manner: The secondary winding of the power transformer is connected to a rectifier tube for fullwave rectification. The rectified output is smoothed with a 2-section RC filter and regulated at 150 volts with a voltage-regulator tube. The output load resistance is not grounded at one end, but at midpoint, so that output voltages of +75 and −75 are obtained with respect to ground. These voltages are used to bias the sections of a dual diode to

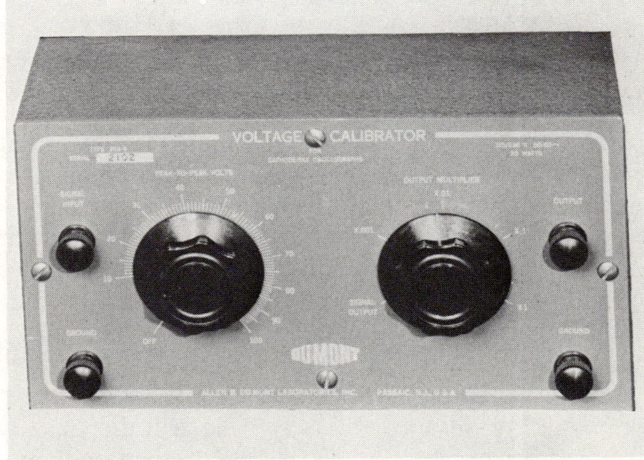

Fig. 7-9. Dumont Type 264-B voltage calibrator.

which a 325-volt ac signal is applied from the transformer secondary. The dual diode clips this signal at +75 and −75 volts, thus giving essentially a square-wave signal of 150 volts. This signal is dropped to 100 volts peak-to-peak by a series divider, and the 100-volt signal is applied to the decade attenuator of the calibrator. A continuously variable control adjusts the attenuator output from 0 to maximum as indicated by a calibrated dial. The dial scale from 0 to 10 is not subdivided in order to avoid inaccuracies that might arise at this end of the scale.

Since the operation of any one of these units is similar to that of the others, an outline of operating procedure is given for only one, the Simpson Model 276. This outline is shown in Fig. 7-10.

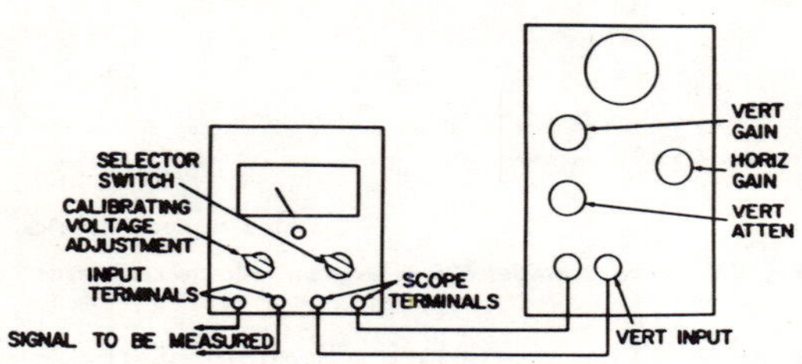

STEP 1 Connect scope, scope calibrator, and signal source as shown.

STEP 2 Adjust HORIZ GAIN to zero.

STEP 3 Turn SELECTOR SWITCH for direct signal feed through.

STEP 4 Adjust scope VERT GAIN and VERT ATTEN for convenient deflection of unknown signal.

STEP 5 Turn calibrator selector switch to appropriate position, and adjust CALIBRATING VOLTAGE ADJUSTMENT to obtain same deflection as in Step 4.

STEP 6 Read voltage of applied signal on meter scale which corresponds to selector-switch position.

Fig. 7-10. Procedure for measuring amplitude of waveform with scope and external calibrator.

The following hints may help in obtaining the best results with voltage calibrators. It is well to remember that when a signal is fed through the calibrator to the oscilloscope input, the result is not quite the same as when direct connection is made to the scope itself, because a small amount of distributed capacity is present in the wiring of the calibrator and is placed in shunt across the oscilloscope input. The capacitance will not be high—one calibrator is listed at 20 pF, and another, at 70 pF—but this might affect some complex waveforms in high-impedance circuits. The operator can avoid this possibility, in special cases, if he will dispense with the handy feature of direct feedthrough for the moment and connect the scope directly to the signal takeoff point. Then, after the operator has set the oscilloscope controls for a certain size of waveform, he can reconnect to the calibrator for voltage measurement. If necessary, a low-capacitance probe could be used for minimum loading and distortion.

Another precaution to observe is that the scope amplifiers are not overloaded at any time during the calibration. The operator can easily identify the point of overload; it is the point at which the waveshape changes while the controls are adjusted for size.

If the calibrator output is a square wave or clipped sine wave of low frequency, there may be a slight tilting of waveshape with some scopes because of phase shift. This can be checked by expanding the trace horizontally. Such a tilted waveform might appear as in Fig. 7-11. In this illustration, the distance *h* is the true measure of the height of the calibrating waveform and is the portion that should be used for reference. If the horizontal deflection is reduced to zero, the waveform appears as a vertical line, shown at the right of the figure. Using the entire length of this vertical trace for calibration reference woold result in a voltage indication slightly less than the true value. Therefore, this check should be made for possible tilt before setting the vertical reference trace.

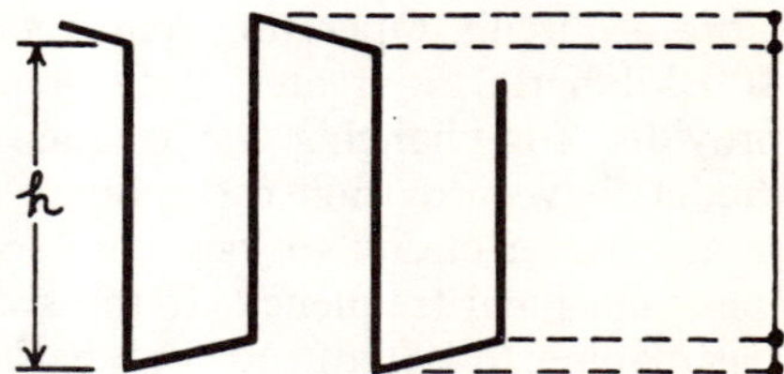

Fig. 7-11. Correct reference height of calibrating voltage.

ELECTRONIC SWITCHES

The electronic switch makes it possible to display two waveforms simultaneously on a single oscilloscope screen. Certain observations can thus be made more easily or conveniently than would be possible without its use. Comparisons can be made between signals at different circuit points to determine the effect of the intervening circuits with respect to phase, frequency, or amplitude differences. Distortion occurring between two points can be observed. The electronic switch is especially effective when one wishes to determine the results of circuit changes or adjustments while the adjustments are actually being performed. Another means for obtaining dual traces is through the use of a double-beam, cathode-ray tube. This method is not commonly encountered, usually being limited to laboratory or special-purpose oscilloscopes. The electronic switch does not require any special provisions for use and will work with any general-purpose scope. As with other types of equipment, models may differ in the number of features offered.

Theory of Operation

The basic principle of operation of most electronic switches consists of applying the two signals to be observed to two separate amplifier channels in the switching unit and at the same time applying a square-wave signal to the amplifiers so that each is alternately driven to cutoff. The output of each channel is developed across a common load (which can be either in the plate or cathode circuit) so that first one and then the other signal appears at the output terminals. The frequency of the applied square-wave signal is called the switching frequency and can be varied to a greater or lesser extent, depending on the complexity of the instrument.

A simplified block diagram of an electronic switch is shown in Fig. 7-12. The separate channels for inputs 1 and 2 may consist of a single stage for each, or they may have such added refinements as additional stages of amplification preceding the output. Some electronic switches may have cathode-follower inputs for minimum loading effects when connected to circuits. The cutoff signal for the two amplifier channels is usually obtained from some form of multivibrator circuit. Operating frequencies may vary from 20 hertz to 100 kilohertz or greater; both step and vernier adjustments are provided for changing the frequency of the cutoff signal. Some models have provision for applying an external sync signal to the multivibrator circuit to lock it to some multiple or submultiple of the sync-signal frequency. If the switching rate is synchronized in this manner to a frequency one-half that of the oscilloscope sweep, the two input signals will appear on alternate traces of the beam across the scope. In this instance, switching is accomplished during beam retrace and is less noticeable than it would be otherwise.

Another method of synchronization is used in most applications. The synchronizing signal is applied to the oscilloscope rather than to the switching unit. This signal can be obtained from one or the

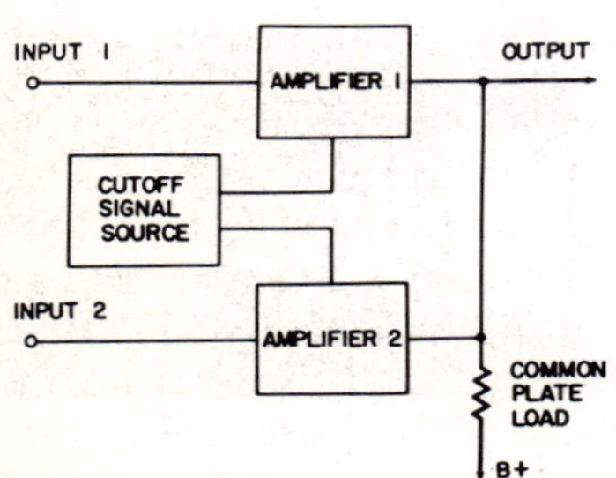

Fig. 7-12. Block diagram of an electronic-switch circuit.

other of the input signals, and the switching frequency is then adjusted to give the least confusion when viewing the two traces. The synchronizing signal is applied to the EXT SYNC jack of the oscilloscope; it is generally undersirable to use internal sync for two reasons. If internal sync is used, the oscilloscope attempts to synchronize with both input signals to the switch and with the switching waveform itself. This makes synchronization uncertain, and a distorted pattern usually results. The second reason is that true phase relationship between input signals will not be maintained. High switching rates have the disadvantage of requiring an oscilloscope of wide response in order to secure the best waveform, because the input signals are superimposed on a signal that is essentially a square wave. It is commonly known that a square wave of high frequency requires a wide-band amplifier for accurate reproduction.

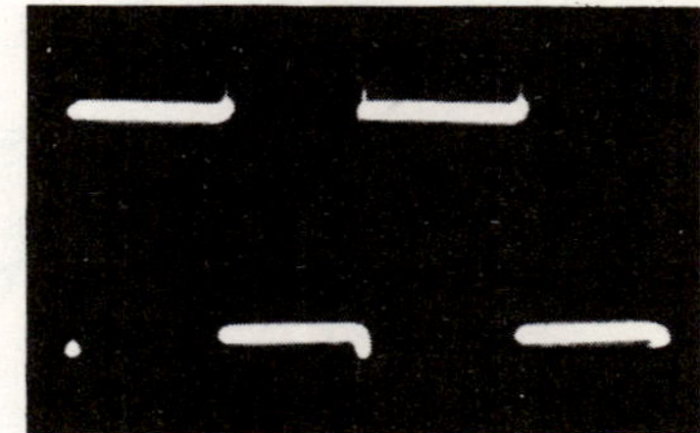

Fig. 7-13. Output signal from the electronic switch with both amplifiers set for zero gain.

If the gains of amplifiers 1 and 2 are reduced to zero, the output of the electronic switch will be a square wave (as shown in Fig. 7-13). As the gains of the amplifiers are increased, the two input signals will appear superimposed on the upper and lower portions of the square wave. Proper selection of the switching frequency will cause both traces to appear continuous, one above the other. A control, variously called BALANCE, AXIS SHIFT, POSITIONING, or the like, provides for separation or merging of the two traces on base lines. This is illustrated in Fig. 7-14. Thus, two waveforms can be displayed, one directly above the other, or they can be caused to merge into one trace for more exact comparison.

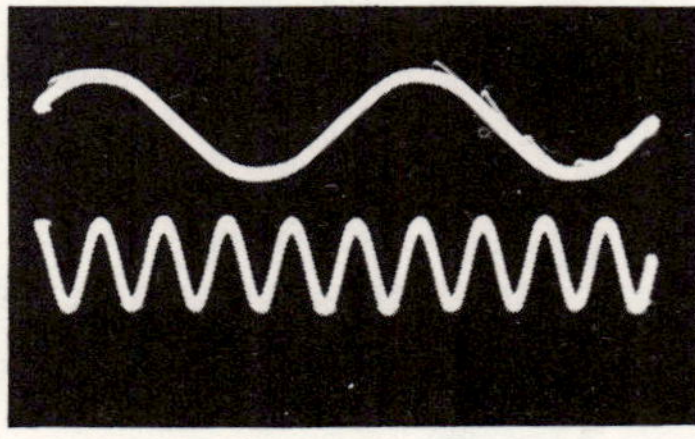

(A) Axis separated.

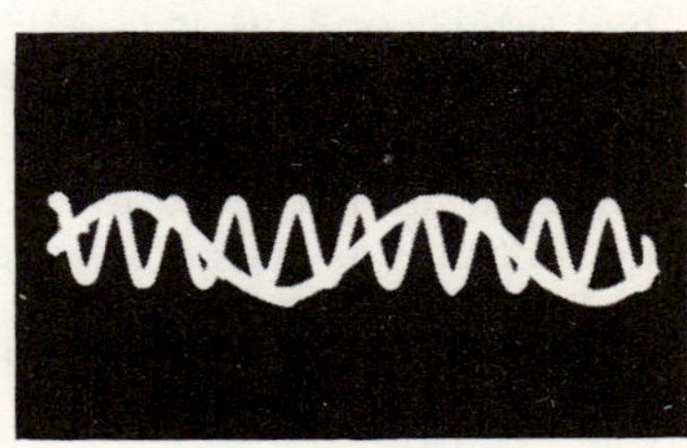

(B) Axis merged.

Fig. 7-14. Effect of axis shift.

Multiple Signals

By using as many switching units as necessary, the technician can display a greater number of waveforms, until practical limits are reached. Fig. 7-15 is a block diagram showing a setup for viewing four signals simultaneously on a single oscilloscope screen. Other combinations can be easily visualized. The waveform shown in Fig. 7-13 indicates that the electronic switch can also be used as a square-wave generator, and this is verified by the operating manual for the unit. Compared to the capabilities of a normal square-wave generator, its use in this capacity may be somewhat limited, but a worthwhile range of frequencies and amplitude variations are usually covered. The use of square waves for amplifier testing will be discussed in Chapter 10. Practical applications for the electronic switch will also be described in subsequent chapters.

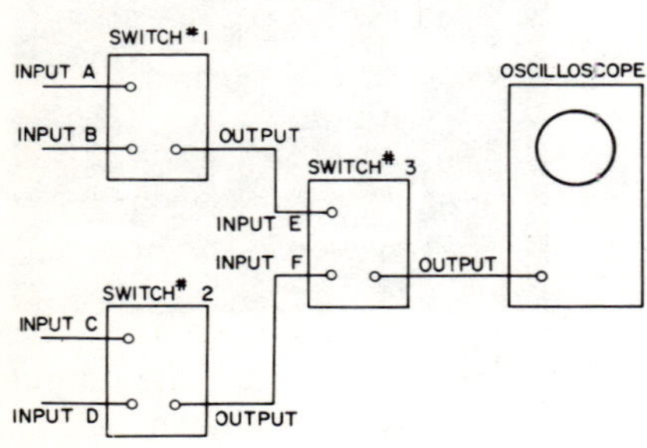

Fig. 7-15. Electronic switch arrangement for viewing four signals simultaneously on an oscilloscope.

An electronic switch of laboratory caliber is pictured in Fig. 7-16. This unit has two amplifier channels, A and B. Each channel can be used in either an ac or dc manner and is supplied with nine attenuation ratios. The frequency response of either channel has the following characteristics:

Direct coupling—Flat to direct current. With 60 pF external load, response down not more than 1 dB at 10 MHz or 3 dB at 15 MHz. With 100 pF external load, response down no more than 1 dB at 6 MHz or 3 dB at 11 MHz.

Capacitive coupling—Down no more than 30% at 5 hertz. High-frequency response identical to direct coupling.

Coarse and fine separation controls are provided. The switching rate control is a 4-position switch. In one position a triggering signal can be applied from an external source to trigger the switching circuits at rates of 0 to 100 kHz. At the other three positions the switching circuits are free-running at rates of either 1 kHz, 10 kHz, or 100 kHz. During triggered operation, a choice of either positive or negative trigger polarity can be had. Sensitivity to the triggering

Courtesy DuMont Laboratories, Div. Fairchild Camera and Instrument Corp.

Fig. 7-16. The Dumont Model 330 electronic switch.

signal is adjustable by a control. The circuit design provides for direct coupling to the dc input of an oscilloscope.

GENERATORS

Generators are associated so often with oscilloscopes that they are mentioned here even though they may not be considered accessories. The square-wave generator has already received some mention, and in addition there are audio, rf, and sweep generators to be considered.

The audio generator furnishes a sinusoidal signal covering the audio-frequency band. The lower frequency limit may be around 10 to 30 hertz, while the upper limit may be as high as 100 or 200 kHz, although the upper frequency limit of hearing for the human ear may be anywhere from 12,000 to 20,000 hertz. The audio generator signal is useful when checking amplifiers for response, amplification, distortion, output power, phase shift, and so on.

The rf or radio-frequency generator supplies a signal covering the band of radio frequencies, although a single generator will not cover the entire band, including as it does rf, vhf, and uhf. These generators are useful for alignment, signal tracing, frequency measurement, response curve marking, and other applications.

A sweep generator is merely an rf generator (or in rare cases, an audio generator) whose frequency is made to vary back and forth above and below a center frequency determined by the dial setting.

The rate the frequency varies is usually 60 times per second, but at one time 400 cycles per second was a popular rate in fm sweep generators (some people called them wobbulators at that time, too). A common way of generating the sweep signal is to feed a variable control voltage to some reactance element of the rf oscillator stage. This voltage causes the oscillator frequency to vary above and below center.

At one time, mechanical means for sweep generating were popular. These usually took the form of a motor-driven capacitor. The capacitor was a part of the tuned oscillator circuit, and as it was made to vary in capacity by the motor, the oscillator frequency varied with it.

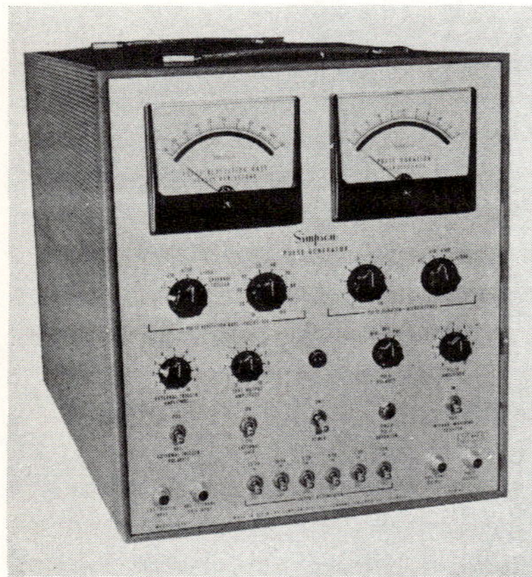

Courtesy Simpson Electric Co.

Fig. 7-17. A lab-type pulse generator.

Another mechanical system is used in some present-day generators. A metal disc is mounted near an inductor that is part of the tuned oscillator circuit. The disc is driven by an attached coil in much the same manner as a speaker diaphragm is driven by its voice coil. The vibrations of the disc cause the inductance to vary, and the oscillator frequency varies in like manner.

Sweep generators, together with an oscilloscope, furnish a convenient means for developing a response curve of tuned circuits or other frequency-discriminating circuits. The response characteristic of an entire video i-f strip of a tv receiver can be developed, and the effects of any alignment adjustments can be seen immediately.

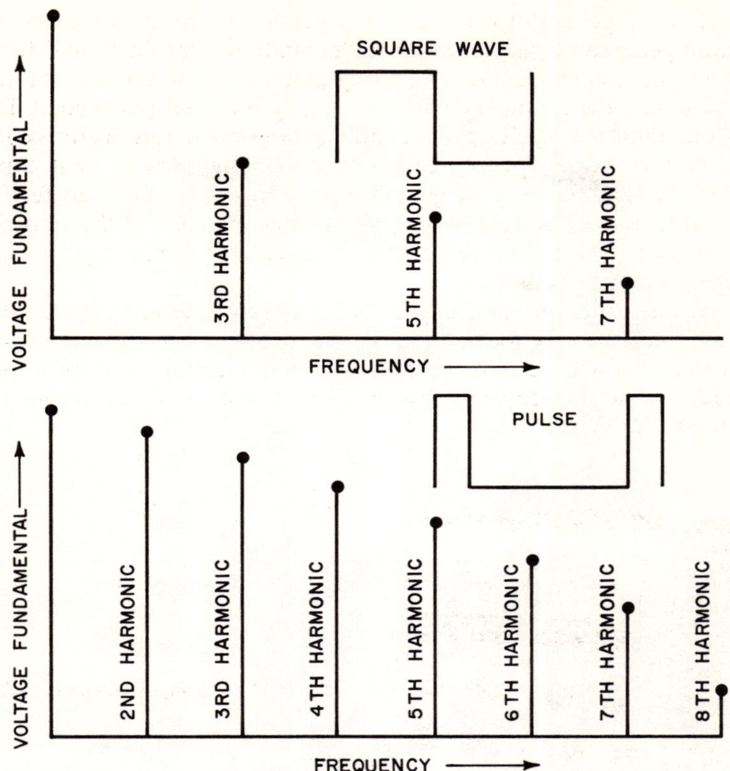

Fig. 7-18. Comparison of harmonic amplitude in square waves and pulses.

Brief mention should also be made in this regard concerning pulse generators, such as illustrated in Fig. 7-17. Pulse generators are used with scopes to check the *transient* response of video amplifiers. When used in combination with a marker generator or signal generator, a pulse generator also shows the transient response of the rf, i-f, and video sections of a television receiver. This is the most exacting test that can be made of a receiver. A pulse generator provides a better test of high-frequency characteristics than a square-wave generator because the harmonics in a pulse have comparatively large amplitude, as shown in Fig. 7-18. Transient testing of receiver sections, circuits, and components is an extensive subject and cannot be described in these pages. Interested readers are referred to specialized books on pulse generators and testing techniques.

Chapter 8

Adjusting and Servicing the Oscilloscope

A number of oscilloscope control adjustments cannot be considered part of the normal operating procedure. A few of these controls are accessible from the outside of the case—front panel, sides, or rear panel—and the rest are located internally. If a control cannot be reached without removing the oscilloscope from its case, then it probably is fairly stable in adjustment and will rarely need attention, if at all. Adjustment may be made necessary by aging or replacement of tubes and components.

When making adjustments with the instrument outside its case, the technician should use reasonable caution to avoid contact with high voltage sections of the instrument. Potentials as high as 3000 volts are used to operate the cathode-ray tube in some oscilloscopes. The operator's manual for the instrument will usually contain instructions for any adjustment the manufacturer considers advisable for the service technician to make. Some of the adjustments to be found in a number of these manuals are mentioned in the following paragraphs.

DC BALANCE CONTROLS

Oscilloscopes having dc inputs or amplifiers dc–coupled from the vernier gain control onward to the deflection plates will usually have a dc balance control. The reason such a control may be needed can be seen by examining Fig. 8-1, which is a simplified version of the first stages of the horizontal amplifier of an oscilloscope.

The cathode (pin 3) of V101A is directly connected to the grid (pin 6) of V102A through R111 and R119. R111 is the horizontal gain control. As this control is varied from one extreme to the other, any ac signal across R111 will be applied to pin 6 of V102A in either increasing or decreasing strength. At the same time, if there is any dc potential across R111, the dc voltage applied to pin 6 of V102A will also change, causing a shift of position in the trace on the screen. The following test will determine whether a dc balance adjustment is necessary. Rotate the gain control from one extreme to the other. If the trace shifts its position, a balance adjustment is needed.

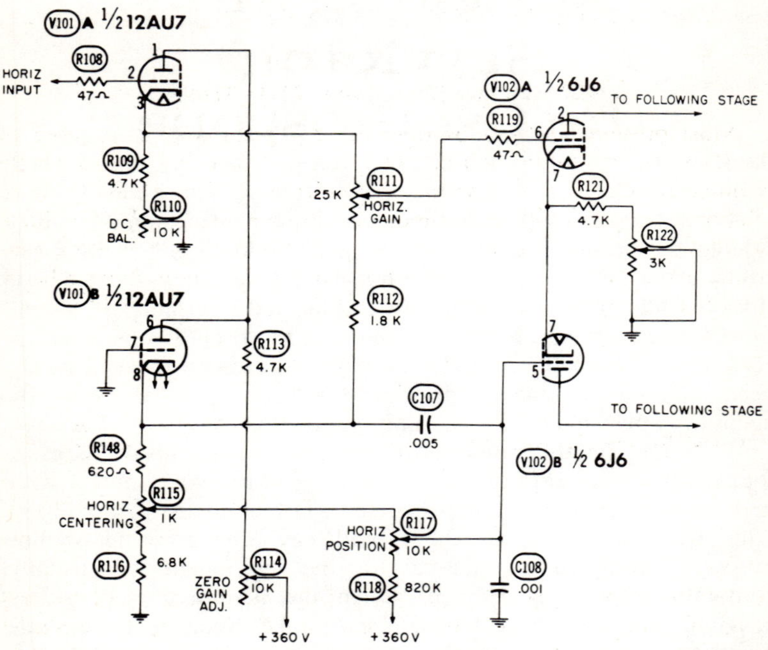

Fig. 8-1. Input stage of an oscilloscope having dc balance adjustment.

Fig. 8-1 shows the dc balance control, R110, in series with the cathode resistor, R109. As R110 is varied, the dc potential of pin 3, V101A, also varies. When pins 3 and 8 of V101 are at the same potential, there will be, of course, no voltage drop across the horizontal gain control, R111, and rotation of this control will not affect the position of the trace. Adjustment of R111 is easily done, then, by the trial method: R110 is rotated, a little at a time, until rotation of the horizontal gain control has no effect on beam positioning. Direct-current balancing of the vertical amplifier is similarly made. Other controls, sometimes called dc balance controls, are used to ensure balanced operating conditions between symmetrical halves of push-pull circuits.

Another adjustment that can be explained using the diagram of Fig. 8-1 is that of the horizontal centering control (R115 in Fig. 8-1). The purpose of this adjustment is to ensure that the beam will be centered horizontally when the horizontal positioning control is turned to the midpoint of its range. For the adjustment, R117 is first set to the midpoint of its range; then R115 is adjusted so that the beam strikes the screen at the horizontal center. The beam is vertically centered by a similar adjustment in the vertical amplifier of the oscilloscope.

VOLTAGE CALIBRATION ADJUSTMENT

Most present-day oscilloscopes are equipped with some type of built-in voltage calibration circuit, even if it provides only a single voltage brought out to a front-panel terminal. Those having more elaborate systems will sometimes provide for adjusting the voltage applied to the calibration circuits. A simple voltage dividing network like that of Fig. 8-2, connected across a winding of the power transformer, can be used.

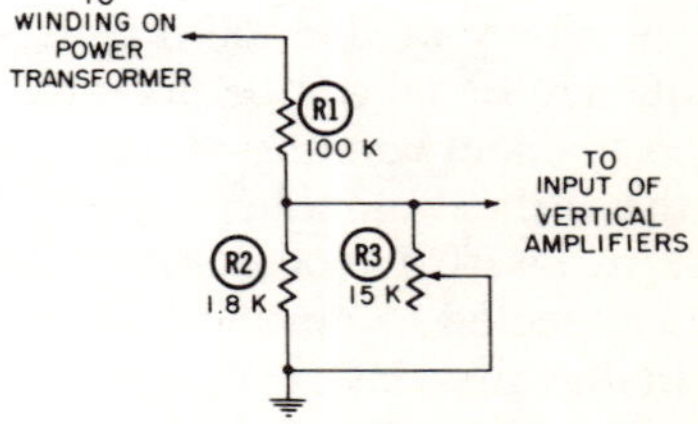

Fig. 8-2. Network for adjusting calibrating voltage.

In this example, R3, the 15K-ohm control, is adjusted to provide .1-volt peak-to-peak for calibration of the vertical amplifier. In another example, the network was connected to one of the low voltage rectifier plate windings of the power transformer and adjusted to give a basic calibration voltage system, using a regulator tube to stabilize the calibration voltage. When adjustments of this sort are made, the power line voltage to the oscilloscope should preferably be applied through a voltage adjusting transformer so that the oscilloscope can be operated at the voltage for which it was designed.

VERTICAL AND HORIZONTAL TV SWEEPS

These sweeps are provided at separate positions of the coarse frequency control so that the operator may select them with a minimum of trouble. They are set up to operate at 30 hertz and 7875 hertz so that two cycles each of the vertical and horizontal waveforms are displayed. Since they may wander in frequency slightly over a period of time, adjustments are provided for resetting them to their original frequency.

The operator's manual for a certain oscilloscope recommends the following calibration procedure:

1. Set the coarse frequency control to H (tv).
2. Connect an audio generator to the vertical input binding post and set to a frequency of 7875 hertz.
3. Set the locking or sync amplitude control to zero.
4. The sync selector switch should be at the Internal position.
5. Adjust the proper calibration control until the pattern locks in for a one-cycle waveform.

The same procedure would be used for the V (30 hertz) position of the coarse frequency control, except that the generator would be set to 30 hertz and the proper calibration control would be adjusted for a one-cycle pattern. If the technician wishes, he may set the 30-hertz sweep by means of a 60-hertz, line-frequency signal, using Lissajous figures (Chapter 9). This method might be more accurate than the other if there is any reason to doubt the accuracy of the audio generator setting at 30 hertz. The line-frequency signal can be obtained from some point inside the scope, say a filament winding of the transformer, and should be connected to the vertical amplifier input. If connected internally, it might be wise to insert a blocking capacitor, since some filament windings are connected to dc voltage points. The adjustment then proceeds as in the first instance, except that a two-cycle pattern must be obtained. The 60-hertz vertical signal will make two vertical excursions while the sweep is making one horizontal excursion. The waveform will look like an infinity symbol.

Note that these adjustment procedures recommend setting the sweep frequencies at 30 and 7875 hertz, or exactly half the vertical and horizontal sweep rates of a tv receiver. Other manufacturers may recommend rates a little lower; for example, one operator's manual gives a figure of 25 hertz for the vertical and 6000 hertz for the horizontal sweep rates. In Chapter 4 it was mentioned that the application of a synchronization signal tends to increase the sweep frequency of the scope slightly above its natural free-running frequency, so this latter practice may widen the range of stable synchronization for the tv sweep feature. However, by actual test on an oscilloscope, a signal slightly below the free-running frequency of the oscilloscope sweep can be synchronized. Note also that in this discussion of sync stability we compare signal and sweep frequencies having a 1 to 1 ratio, but the same principle holds true where the signal is twice, three times, or some other whole multiple of the sweep frequency. These two positions of the sweep frequency control (H and V, tv) can also be calibrated by using a video signal from a tv receiver. The 7875-hertz sweep is set for two horizontal lines of video pattern, and the 30-hertz sweep for two fields.

COMPENSATED ATTENUATORS

The circuit construction and theory of compensated input attenuators were shown in Chapter 5. The adjustment procedure will be discussed here. It applies equally well to the adjustment of the low-capacity probe to be used with the oscilloscope. There are two major requirements for performing the adjustment—a suitable input signal and some indicating device to show the effects of adjustment. The oscilloscope screen can be used as the indicating device, and a square-wave generator will provide the signal. The trimmer for each attenuator position is adjusted for the best square-wave pattern on the oscilloscope screen. There should be no rounding or peaking of the leading edge of the square wave. Fig. 8-3 shows the response for different degrees of adjustment of the trimmers. In Fig. 8-3B the trimmer capacity is too small, in Fig. 8-3C, it is just right, and in Fig. 8-3D it is too large.

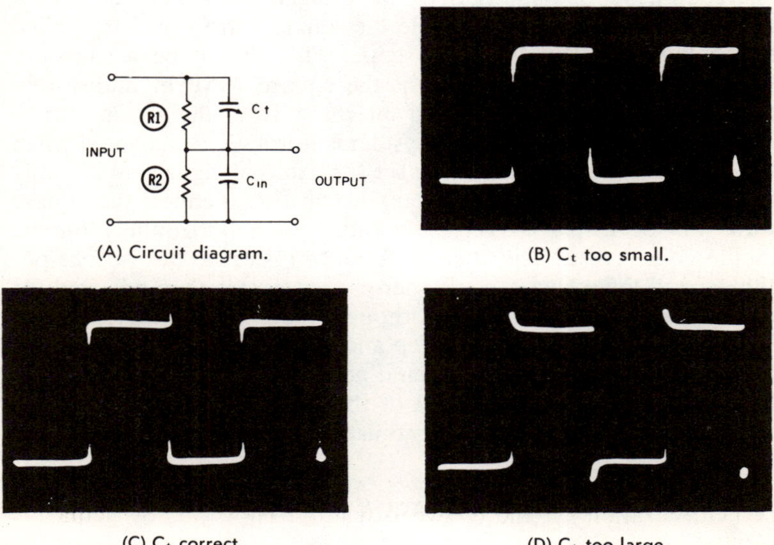

(A) Circuit diagram.

(B) C_t too small.

(C) C_t correct.

(D) C_t too large.

Fig. 8-3. A frequency compensated attenuator and the effect of various degrees of adjustment on a square-wave signal.

The frequency selected for the square-wave signal is important. Usually, a frequency of about 10 kilohertz per second is satisfactory. Since the attenuator network and the amplifiers are in series, one of them will be a limiting factor in the total response of the combination. Theoretically, the attenuator network will attenuate equally all the frequencies applied to it if it is properly adjusted, whereas the amplifier response of many general-purpose scopes will start to fall off around 3 or 4 megahertz or before. The amplifier, then, is the weak link in the response chain, and it should be adjusted first if adjustment is required.

Since the amplifier is the limiting factor in the response of the vertical system, the square-wave frequency should be chosen to suit its requirements. That is, there would be no logic in trying to adjust the attenuator with a signal that would not be passed by the amplifiers. There is a rule of thumb stating that good response to a square-wave signal indicates good response to a sine-wave signal of frequencies ranging from 1/10 to 10 times the square-wave frequency. Following this rule, a 100-kilohertz square wave is suitable for testing an amplifier to frequencies of 1 megahertz. Actually, whether a 10 or a 100-kHz signal is used will make little difference in the final adjustment of the attenuator.

Attenuators in the horizontal amplifier circuit can be adjusted in much the same manner as the vertical attenuators, but there is one complication—an external sweep signal must be provided to drive the vertical amplifiers while the square-wave signal is applied to the horizontal input of the scope. This should be a sawtooth sweep signal in order to present the square wave in undistorted form, but it cannot be obtained internally from the sweep system of the scope itself because this system is normally deactivated when the horizontal selector switch is set for external signals. With sawtooth and square-wave signals applied in this manner, the square wave will be displayed vertically rather than horizontally on the oscilloscope screen. The attenuator control is set to the various positions, and the associated trimmer capacitors are then adjusted for best square-wave response. Although there is no provision for synchronizing the sawtooth sweep and the square-wave signal, they can be kept almost in step by hand adjustment of one generator or the other. A slow rate of travel of the waveform across the scope face will not interfere with the adjustment procedure.

LOW-CAPACITY PROBES

These probes can be adjusted for frequency compensation in the manner just described for vertical attenuators. A square-wave frequency of 10 kHz should be satisfactory in most cases. If the vertical input attenuator of the oscilloscope is set to the X1 position, the signal will feed straight through the attenuator, thus avoiding any effect the attenuator might have.

In the absence of a square-wave generator, the technician can adjust his probe by using the video signal obtained at some point in a tv receiver. The horizontal and vertical sync pulses of this signal have the characteristics of square waves and, because they do differ widely in frequency, are suitable for this method. The video signal should be fed through the probe to the vertical input of the oscilloscope. The oscilloscope is synchronized to the vertical sync pulses, and the waveform will appear as in Fig. 8-4. As the

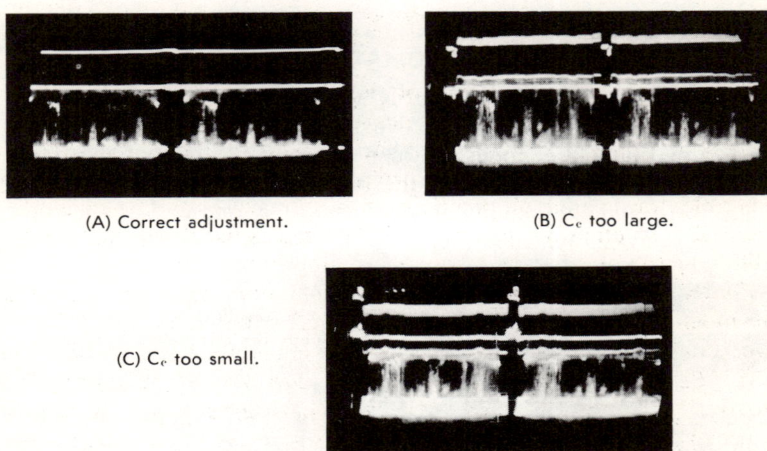

(A) Correct adjustment.

(B) C_c too large.

(C) C_c too small.

Fig. 8-4. Video signal used in adjusting a low-capacitance probe.

trimmer is adjusted, the level of the horizontal sync pulses will shift with respect to the vertical sync pulse level. The correct adjustment is obtained when both the horizontal and vertical sync pulses are on the same level, as in Fig. 8-4A. This method may not be as accurate as the first one, but it is practical when the technician does not have access to a square-wave generator.

FREQUENCY RESPONSE OF THE AMPLIFIERS

The frequency response of wide-band oscilloscopes is obtained through careful design directed at bringing up the response at the extreme high- and low-frequency ends of the band. Because of natural factors, the response normally falls off at these extremes. Filter and coupling capacitors tend to limit the response at the low end, while shunt capacities limit the response at the high end. Several methods of extending the frequency response of oscilloscope amplifiers are described in Chapter 5. These methods include the use of series and shunt peaking coils, and feedback circuits. Some of these components are adjustable in some oscilloscopes. The manufacturer's instructions for adjusting these components should be faithfully observed if adjustment becomes necessary, and haphazard methods should not be tried. For this reason, no specific adjustment procedure is given here. A complete adjustment of a vertical amplifier system with compensated attenuators, as described by one oscilloscope manufacturer, requires the use of another oscilloscope, an rf demodulator probe, a video sweep generator, a video marker source, and a square-wave generator.

ASTIGMATISM

Astigmatism in oscilloscopes is a defect in focusing similar to the same defect in the human eye. Astigmatism is manifested by an unequal degree of focus between different portions of the trace. That is, vertical excursions of the beam may give a sharper trace than horizontal excursions, or vice versa. Or, one end of the trace may be sharper than the other. Fig. 8-5 shows this effect in a trace

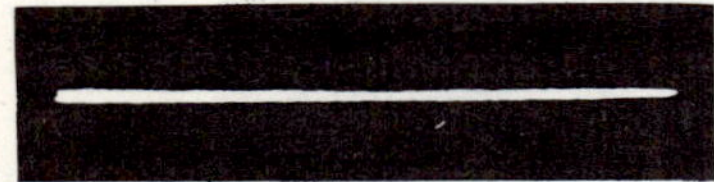

Fig. 8-5. Oscilloscope trace which exhibits the effect of astigmatism.

focused at the extreme right but not at the left. When a corrective adjustment is provided, its use is fairly simple. First, the oscilloscope controls are set to obtain a trace across the cathode-ray tube face (a simple waveform serves admirably), and the astigmatism control is adjusted for the best compromise focus throughout the trace. A circular waveform is one of the best for this purpose since it makes every possible change in direction of the trace.

OTHER ADJUSTMENTS

A number of bias adjustments are provided in some oscilloscopes. These adjustments are made to obtain optimum operation of some of the more important or more critical stages. Linearity of sweep is important in obtaining an accurate waveform, and controls are sometimes provided so that the sweep may be made as linear as possible. Another oscilloscope is the best device for checking sweep linearity. The sawtooth waveform should have the straight sides indicated in Fig. 3-1. Probably the next best test for sweep linearity, if a second oscilloscope is not available, is to synchronize a number of sine waves on the scope screen and note whether the waveform appears compressed or expanded at either end. Sweep linearity is difficult to judge by observing a single cycle of a sinusoidal waveform, but if a number of cycles are viewed at once, crowding or expansion at one end of the sweep is easier to see.

If an oscilloscope has a driven sweep, an adjustment for controlling the operation of this feature is usually provided.

SERVICING THE OSCILLOSCOPE

The oscilloscope is built with the same type of electronic components and circuit design that the service technician sees daily in his work with receivers and amplifiers. Its circuits function in agreement with the well-known ac and dc laws. Consequently, the same servicing procedures should work equally well with oscilloscopes as with the other electronic equipment that he services. One of the best assets the technician can have is a good knowledge of the construction and function of the different parts of the oscilloscope. His work is made a lot easier, too, if he has another oscilloscope he can use to check the defective oscilloscope. A meter and a tube tester should also be added to the list of necessary test equipment. A schematic of the oscilloscope, with proper voltages and resistances indicated at important points, will be a great help.

Oscilloscope troubles might be grouped roughly under two classifications: (1) those that appear suddenly, and (2) those that suggest gradual deterioration over a period of time. The latter class of troubles may even escape notice unless the performance of the oscilloscope is checked. Performance checks will be discussed in subsequent paragraphs.

The technician can sometimes obtain an idea of the location and cause of the trouble by noting the effect when the controls are operated. Any abnormal effect may be the clue to start him on the correct line of reasoning. If no direct clue is obtained by working the controls, the technician might start by testing the tubes. If the cathode-ray tube produces a spot or trace that responds to positioning, focus, and intensity controls, it would probably be passed by a tube tester. Gas triode sweep oscillator tubes like the 884 may test all right in a tube tester and yet not perform satisfactorily in actual service; so the final test here should be direct substitution.

The nature of the complaint will often suggest which section of the oscilloscope to examine for the defect. Thus, poor synchronization will most likely result from a fault somewhere between the sync takeoff point and the sweep oscillator; poor focusing, positioning, or intensity suggests either the controls themselves or the voltage network supplying the controls; reduced sweep suggests trouble in either the sweep oscillator or the horizontal amplifier; and so on.

In any servicing procedure, the technician should use the same precautions against shock that were recommended earlier in the chapter.

Chart 8-1 is designed to aid in troubleshooting general-purpose oscilloscopes. Once a trouble has been localized to a certain point or stage, the defective component can probably be found by visual

Chart 8-1. Oscilloscope Troubleshooting Chart

Trouble	Possible Remedy
Pilot light will not operate	Check ac power cable, fuse, pilot light, or power transformer.
No spot on cathode-ray tube	Check horizontal or vertical positioning control, high voltage, vertical and horizontal amplifier plate load resistor.
No control of focus or intensity	Check cathode-ray tube socket, high voltage bleeder open, focus control, and intensity control.
No control of vertical positioning	Check vertical amplifier, vertical positioning control, low-voltage power supply, and vertical amplifier tubes.
No control of horizontal positioning	Check horizontal amplifier, horizontal positioning control, low-voltage power supply, and horizontal amplifier tubes.
No vertical deflection	Check vertical amplifier tubes, low-voltage power supply, input leads, and vertical attenuator switch.
Vertical deflection but poor frequency response	Check input and output stage frequency compensators.
No horizontal sweep on some settings	Check coarse frequency range switch, and range capacitors.
No horizontal sweep on any setting	Check sweep circuit oscillator tube.
Horizontal deflection but poor frequency response	Check horizontal amplifier tubes, attenuator switch, low-voltage power supply, input leads, and output frequency compensators.
No sync	Check sync selector switch, sync injector tube, and locking control.
No blanking	Check coupling capacitor.

inspection—as in the case of a burnt resistor—or by voltage and resistance checks. Signal tracing methods, using an audio generator and another oscilloscope, can be resorted to in stubborn cases.

CHECKING PERFORMANCE

Performance checks might well start with a test of all tubes to make sure that the substandard performance is not due to a weak tube or two. Here are some tests the scope owner can apply to check the condition of his instrument. A source of ac sine-wave signal is needed for some of the tests. This source can be an audio generator, or where a 60-hertz signal is satisfactory, the test-signal source supplied at a binding post on some scopes may be used.

Vertical Amplifier Sensitivity

Two methods for checking the sensitivity of the vertical amplifier can be used. The easier and more accurate method—the use of a scope calibrator—will be discussed first.

Connect the calibrator to the vertical-input terminals and set the vertical amplifier control and input attenuator for maximum sensitivity. Adjust the calibrator for a vertical deflection of one or two inches on the scope. Read the input signal from the scope calibrator. Since this reading is in peak-to-peak volts and most scope manuals give the amplifier sensitivity in rms volts per inch, it will be necessary to convert to rms volts. This conversion can be done by dividing peak-to-peak volts by 2.8. This value is then divided by the scope deflection in inches to obtain the sensitivity. For example, assume that a deflection of two inches on the scope is obtained by a 140-millivolt signal from the calibrator. Then:

$$\text{Sensitivity} = \frac{140}{2.8 \times 2}$$
$$= 25 \text{ rms millivolts per inch}$$

The second method of checking vertical amplifier sensitivity entails the use of an audio sine-wave generator and a meter for reading ac rms volts. Since the modern oscilloscope has a high sensitivity, the generator output must be attenuated by means of a simple divider network. A 4700- and 510-ohm resistor across the generator output terminals would allow approximately one-tenth of the generator output to be applied to the scope from across the 510-ohm resistor. Connect an ac vtvm across the divider network. This meter must operate as a vtvm on the ac positions in order to obtain reasonable accuracy. Adjust the generator output for a convenient vertical deflection on the scope, as in the first method.

Then read the rms voltage applied to the divider network. As an example, it might be 500 rms millivolts for a 2-inch deflection. Only one-tenth of the voltage appearing across the network is applied to the scope. The voltage applied here is 50 millivolts. Since the measurement is already in rms volts, no conversion is needed; we merely make the following calculation:

$$\text{Sensitivity} = \frac{50}{2}$$

$$= 25 \text{ rms millivolts per inch}$$

Horizontal Amplifier Sensitivity

Horizontal sensitivity is obtained in much the same manner as the vertical sensitivity, except that the signal must be applied to the horizontal amplifier. The horizontal amplifier sensitivity is usually much less than the vertical sensitivity (that is, it requires a greater input for 1-inch horizontal deflection). The signal most commonly applied to the horizontal amplifier is taken internally from the horizontal-sweep circuit and is of such magnitude that less amplification is necessary.

Vertical Amplifier Frequency Response

One of the fastest methods for checking amplifier response is through the use of square waves. This method requires some source of square-wave signal, such as a square-wave generator. If such a generator is not available, the horizontal sync pulses in a tv video signal provide a good check. It is an accepted rule that good square-wave response indicates good response of the amplifier to frequencies from one-tenth to ten times the fundamental frequency of the square wave. Thus, if an oscilloscope showed acceptable square-wave response as the square-wave frequency was varied from 250 hertz, good scope response would be indicated for the frequency range of from 25 hertz to 2.5 megahertz.

Another method for determining the frequency response of an amplifier, such as the vertical amplifier in a scope, is to take a number of output readings over the frequency range while maintaining a constant input. These readings can then be plotted as a graph to indicate the response. This method is more time consuming than the other one.

Fig. 8-6 shows the response of an oscilloscope to the video-output signal of a monoscope. The oscilloscope was synchronized to show the horizontal sync pulses. The vertical amplifier was set to the 4-MHz bandwidth position. Fig. 8-7 shows the response of the scope to the same signal, but the bandwidth switch was set at the 2-MHz position. The square-wave response is poorer, as

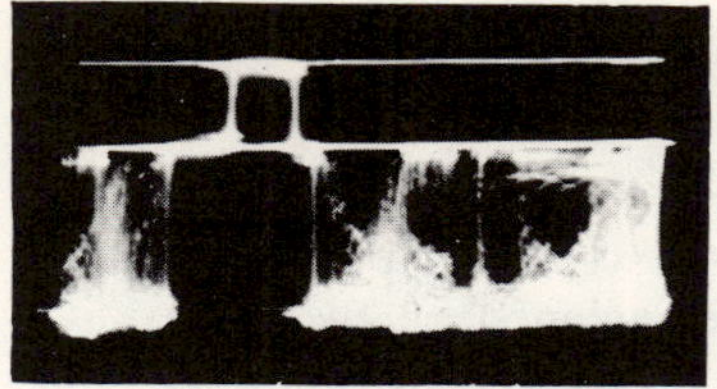

Fig. 8-6. Oscilloscope response using amplifier of 4-megahertz bandwidth.

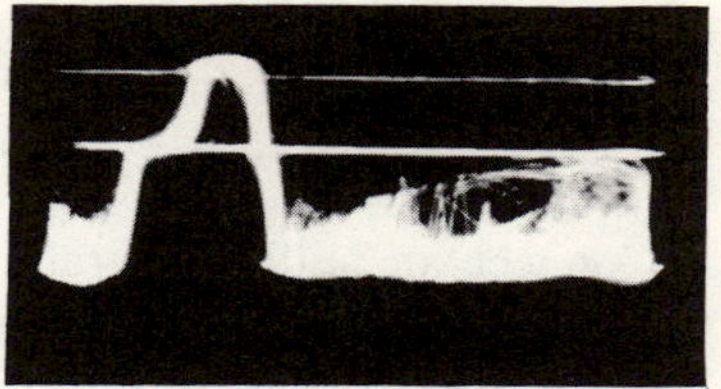

Fig. 8-7. Oscilloscope response using amplifier of 2-megahertz bandwidth.

indicated by the rounded corners. Fig. 8-8 shows the effect of too much input capacity. The scope was left at the 2-MHz position, and an additional capacity of .006 mfd was placed in parallel with the scope input capacity by bridging across the input terminals. As a result, the front porch of the horizontal sync pulse almost disappeared.

Synchronization

Turn the sync amplitude or locking control to zero. Set the sync selector to Internal and apply a moderate signal to the vertical input. Using only the coarse- and the fine-frequency controls, synchronize the signal as nearly as possible. Then advance the sync-amplitude control to lock the signal. If the circuit is operating properly, only a slight adjustment of this control should be necessary.

Frequency Coverage of Sweep

The fine-frequency control or sweep vernier gives a continuous range of sweep frequencies for each setting of the coarse-frequency control. The top frequency of each range should equal or overlap the bottom frequency of the next higher range. The frequency limits of each range can be checked with an audio-signal generator.

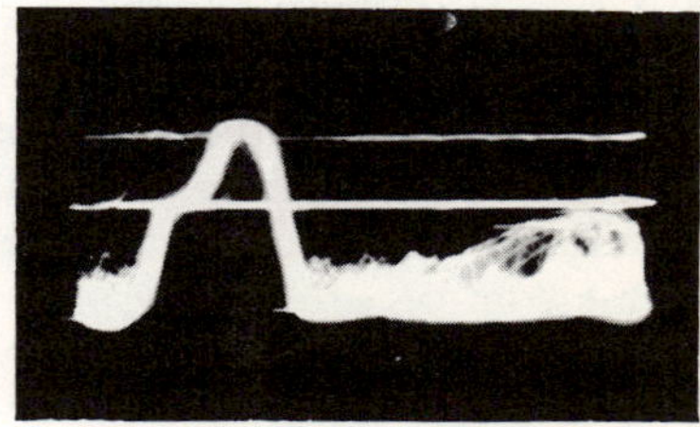

Fig. 8-8. Response showing effect of added input capacitance.

Connect the generator output to the vertical input of the scope. Set the sync-locking control to zero. With the fine-frequency control at each extreme of its range, vary the signal-generator frequency to obtain a stationary pattern of two or three cycles on the scope.

Divide the generator frequency by the number of cycles on the scope pattern to obtain the sweep frequency.

Sweep Linearity

Connect an audio generator to the vertical-input terminals or use the 60-hertz test signal if one is included on the scope. Synchronize the signal with the frequency and locking controls of the scope. Choose a frequency in which several cycles are visible on the screen. Using the horizontal-amplifier control, expand the trace to fill the screen horizontally. The pattern should be evenly spaced throughout its length; if it is crowded or stretched at any portion, nonlinearity of sweep is indicated. If this nonlinearity condition does not disappear as the sweep width is reduced by means of the horizontal-amplifier control, the sweep signal being applied to the horizontal amplifier may be nonlinear, and the cause should be sought in the sweep-generating circuit. On the other hand, if this nonlinearity disappears as the sweep width is reduced, the nonlinearity was not caused by a defect in the sweep circuit, but by some defect in the horizontal amplifier. When some trouble exists in the horizontal amplifier, the amplifier may possibly be overdriven when the sweep width is adjusted to maximum. Maximum sweep width on most scopes extends past the borders of the screen; therefore, the sweep must be moved to either side of the screen with the horizontal-positioning control in order to view the ends of the sweep.

Vertical Linearity

Apply a weak signal to the vertical input. Adjust the vertical gain to minimum and position the resulting spot on the screen to the center of the ruled grid. Advance the vertical-gain control slowly and note whether the signal expands at an equal rate above and below the middle horizontal line of the grid. The dc balance adjustment should be checked, if there is one, to avoid movement of the trace because of misadjustment. Decrease the signal input and increase the scope sensitivity in order to check the extremes of the vertical-amplifier range. Most oscilloscopes incorporate both a variable gain and a step attenuator for the vertical amplifier. These controls are usually placed in the circuit in the order shown in Fig. 8-9. The attenuator appears first, followed by an amplifier stage and the vertical-gain control. Under these circumstances, the first stage can be overloaded by applying an input signal too large for the attenuator position. Then the pattern on the screen would be distorted, no matter how much it might be reduced with the vertical-gain control. Some scopes have the attenuator positions marked with the maximum signal value they are designed to handle.

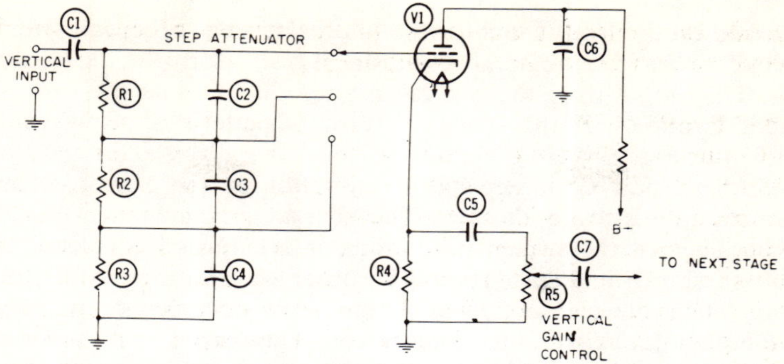

Fig. 8-9. Simplified partial schematic of an oscilloscope showing step attenuator and gain control in vertical amplifier.

This marking helps the operator avoid possible distortion by overloading.

The foregoing paragraphs cover most of the operating controls encountered in the average general-purpose scope. Laboratory scopes and scopes for special applications would, of course, have additional controls.

One of the most frequent calibration procedures for triggered-sweep scopes concerns sweep-speed checks and adjustments. Trimmer capacitors or potentiometers are provided on each step of the sweep-speed control to adjust the sweep speed to the exact microseconds per cm, milliseconds per cm, or seconds per cm, as

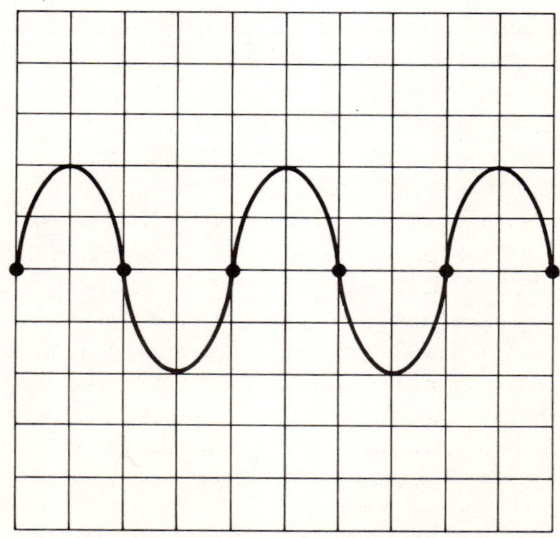

Fig. 8-10. Checking scope sweep timing.

marked on the front panel for the particular step. To check sweep-speed calibration, use an accurate signal generator or audio oscillator. The output from the generator is fed directly into the vertical-input terminals of the scope. In turn, a pattern is displayed as shown in Fig. 8-10.

Suppose that the sweep-speed calibration on the 1 microsecond per cm step is to be checked. The generator might be set at 250 kHz. Then, if the sweep-speed trimmer is adjusted correctly, the pattern shown in Fig. 8-10 will be observed. In other words, one complete cycle of the 250-kHz sine wave occupies 4 cm along the horizontal axis on the scope screen. This means that the sweep speed is 1 microsecond per cm. Other steps on the sweep-speed control are checked in the same basic manner.

Chapter 9

Frequency and Phase Measurements

The oscilloscope is well suited for frequency and phase measurements because it responds instantly and faithfully to input signals applied to it. Frequency measurements will be limited to those frequencies within the response range of the oscilloscope amplifiers. If the signals are applied directly to the deflection plates of the cathode-ray tube, the distributed capacity of the related circuits may be a limiting factor. In either case, the frequency range will extend from low audio frequencies up into the radio frequencies.

One of the simplest (also the least accurate) methods of frequency measurement is with the oscilloscope sweep oscillator. The unknown signal is fed to the vertical input of the scope, and the sweep controls are adjusted for a stationary pattern of one cycle, if possible, or for a pattern of several cycles if the scope sweep rate does not go high enough for a 1 to 1 ratio. The frequency of the unknown signal can then be estimated from this information. This method gives approximate results only because the sweep range switches and vernier controls of most oscilloscopes are not calibrated to any degree of precision. The range switch is usually marked at each position with the frequency limits (low and high) to be covered by a complete rotation of the vernier control. The vernier control is usually indexed by a scale that is evenly divided, but that bears no direct relation to the sweep frequencies obtained. However, if the operator wished to take the time, he could make up a calibration chart for each sweep range position and thus obtain a greater degree of precision. For most accurate results, the sync amplitude control should be set to minimum.

LISSAJOUS FIGURES

More accurate frequency measurements can be obtained with Lissajous figures. Fig. 9-1 shows how these waveforms can be developed. The internal sweep system of the oscilloscope is turned off, and two different sinusoidal signals are fed to the inputs, one to the vertical and the other to the horizontal. The result is a closed loop waveform named after the French scientist Lissajous, who first showed how such figures could be developed, both optically and geometrically.

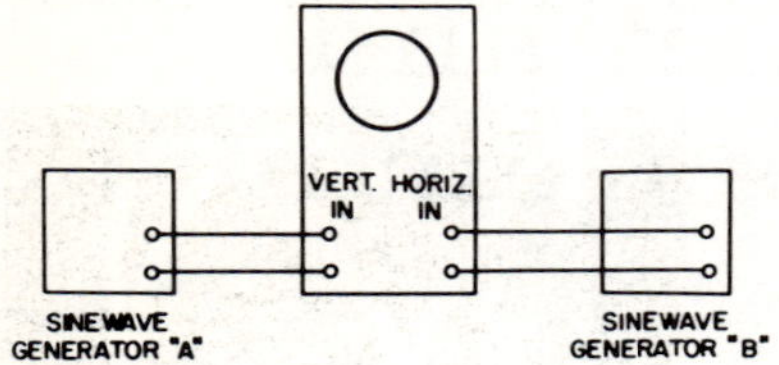

Fig. 9-1. Method for generating Lissajous figures.

If the frequency of one of the generators is known, the other, or unknown, frequency can be determined by proper interpretation of the Lissajous figure generated. The known frequency signal is usually connected to the horizontal input of the scope, and the unknown signal is connected to the vertical input. Simple frequency ratios result in waveforms that are easy to interpret. Therefore, the known frequency generator should be adjusted to obtain such a waveform, if possible. When the two signals have a ratio that can be expressed by whole numbers, such as 2/3, 3/4, 4/1, etc., the waveform will appear to be stationary on the scope screen. If one generator is adjusted to give a signal ratio not quite expressible by two whole numbers, the pattern will seem to rotate slowly in one direction or the other. This rotation appears to take place in three dimensions, a fact that can sometimes be used to advantage in counting the loops of the pattern.

Fig. 9-2 shows the pattern resulting from a 2 to 1 frequency ratio and illustrates the method used for calculating the frequency ratios of other patterns. Note that horizontal line AB is drawn, touching the pattern at two points, and vertical line AC is drawn,

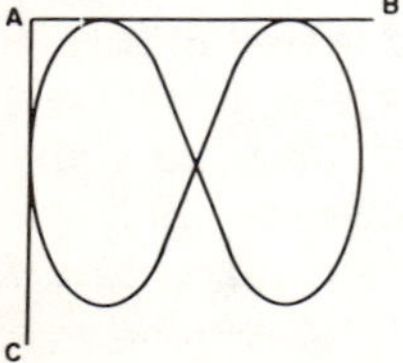

Fig. 9-2. A method for calculating the frequency ratio indicated by a Lissajous figure.

touching the pattern at one point. The points of contact on line AB are caused by vertical excursions of the oscilloscope beam and are therefore related to the frequency of the signal applied to the vertical input of the scope. The points of contact on line AC (only one in this instance) are caused by horizontal excursions of the beam and are therefore related to the frequency of the signal applied to the horizontal input. From the pattern shown in Fig. 9-2, the oscilloscope beam makes two vertical excursions while making one horizontal excursion; therefore, the vertical input signal has twice the frequency of the horizontal input signal. Three points of contact to line AB and one to line AC would indicate that the vertical input signal frequency was three times that of the hori-

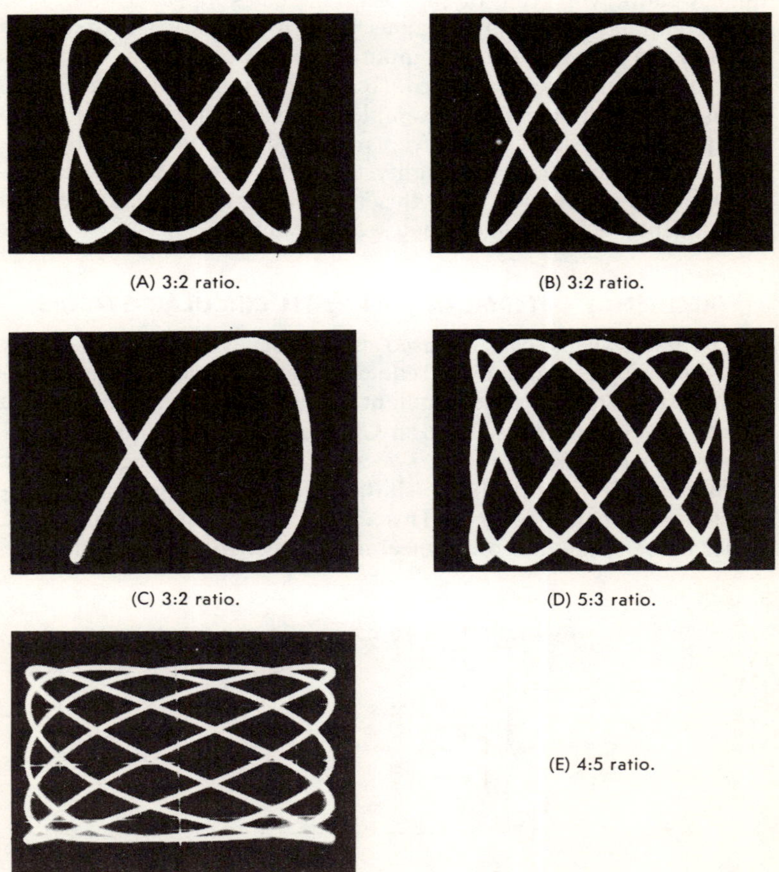

(A) 3:2 ratio.

(B) 3:2 ratio.

(C) 3:2 ratio.

(D) 5:3 ratio.

(E) 4:5 ratio.

Fig. 9-3. Patterns to determine frequency ratios between pairs of input signals.

zontal input signal; three points of contact on line AB and two on line AC would indicate that the vertical input signal frequency was 3/2 that of the horizontal input signal, and so on.

Once the frequency ratio indicated by the Lissajous figure has been determined, the unknown frequency can be found by multiplying the known frequency by this ratio. A few of the simpler ratios are shown in Fig. 9-3. At "A", a 3 to 2 ratio is shown. If this pattern is permitted to rotate, it will appear as shown at B or C. The pattern at C could lead to some error in interpretation. The point to remember is that the pattern should consist of closed loops rather than abruptly terminated single lines as it seems to do in Fig. 9-3C. The pattern of Fig. 9-3D indicates a 5 to 3 frequency ratio, while that of Fig. 9-3E indicates a 4 to 5 ratio.

A continually shifting Lissajous pattern results when the phase relationship between the two input signals is constantly changing. The more complex the pattern (resulting from a frequency ratio having large numbers, for example, 17/13), the harder it is to interpret. The task is made even more difficult by a shifting pattern. It is better, then, to simplify the ratio, if possible, by changing the known frequency. If this is not practical, other methods of frequency determination may work better.

FREQUENCY DETERMINATION WITH CIRCULAR SWEEPS

Circular sweeps can be used in frequency measurements by superimposing the unknown frequency signal on the circular waveform and interpreting the resultant waveform. The circular sweep can be developed as discussed in Chapter 3, and the unknown signal is then impressed on the sweep by placing it either in series or in parallel with the phase-shifting network. A series arrangement is shown in Fig. 9-4. This circuit arrangement results in a pattern resembling a crown wheel in perspective. The crown wheel

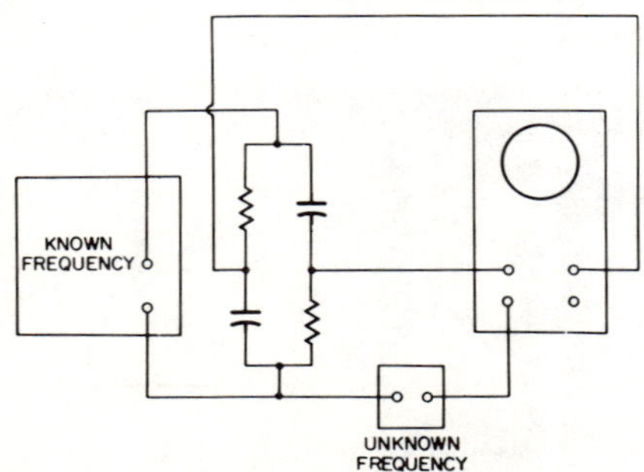

Fig. 9-4. Unknown frequency superimposed on circular trace, resulting in crown wheel pattern.

appears to rotate unless the known signal is adjusted in frequency to obtain a whole number ratio with the unknown signal.

When the known frequency has been adjusted to obtain a stationary pattern, the "cogs" or "teeth" on one side of the crown wheel can be counted and used to calculate the unknown frequency. Fig. 9-5 shows a circular pattern with six teeth. This means the unknown frequency is six times the known frequency. A simple 6 to 1 ratio is indicated by this pattern; the pattern seems to be made up of a single line tracing around the wheel circumference once and outlining six teeth on one edge of the wheel as it does so.

The pattern shown in Fig. 9-6 is a little more involved. Here the pattern appears to be made up of a line going twice around the

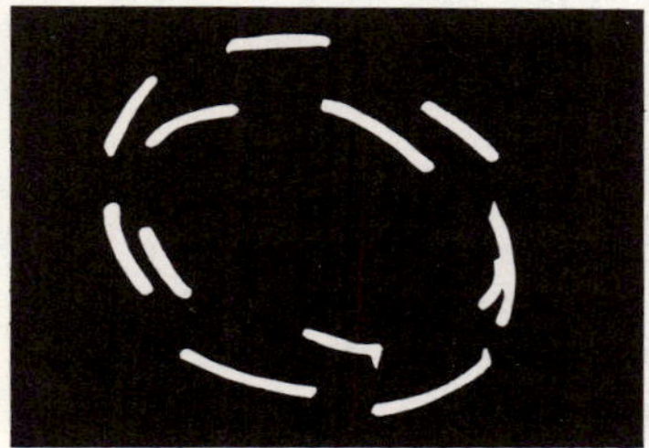

Fig. 9-5. 6-to-1 crown wheel pattern.

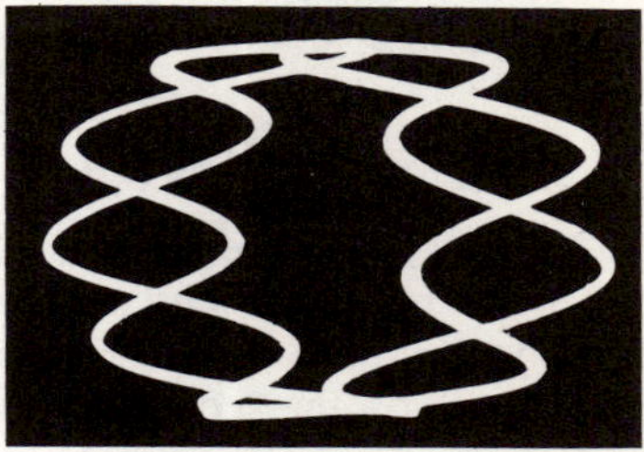

Fig. 9-6. 9-to-2 crown wheel.

circumference, outlining nine teeth as it goes. This represents a 9 to 2 ratio. In other words, the unknown frequency is 9/2 the known frequency (the known frequency signal is used to develop the circular trace).

With this method, as with other methods of frequency measurement, it is best to adjust the known frequency so that simple ratios are obtained, such as 1:1, 2:1, 3:2, etc. Thus, any confusion introduced by more complex patterns will be avoided. Less confusing patterns can be obtained if the unknown signal is used to modulate the circular trace radially as described in Chapter 3.

Lissajous patterns are at their best for frequency and phase measurements when the input signals are sinusoidal or nearly so. Fig. 9-7 shows the resultant pattern when a square-wave signal is applied to the vertical input and a sine-wave signal is applied to the horizontal input of an oscilloscope. The frequency ratio is 1

Fig. 9-7. Pattern that results when a square-wave signal is applied to the vertical input and a sine-wave signal is applied to the horizontal input of an oscilloscope.

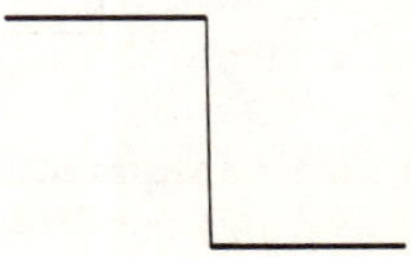

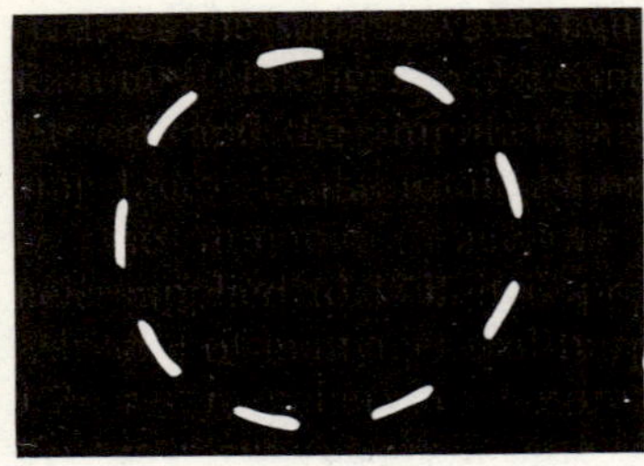

Fig. 9-8. Circular trace with intensity markers for frequency determination.

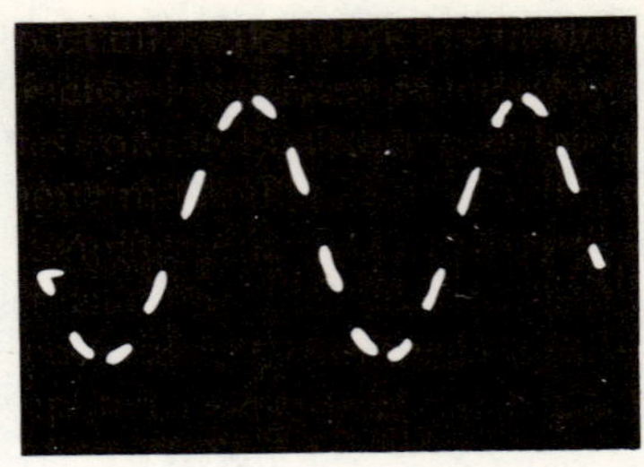

Fig. 9-9. Conventional waveform with intensity markers for frequency determination.

to 1; any other ratio results in a pattern that is practically unintelligible.

MEASURING FREQUENCY BY INTENSITY MARKERS

Unknown frequencies can sometimes be determined by using intensity markers. The unknown frequency is applied to the intensity modulation jack of the oscilloscope to mark the pattern developed on the scope screen by the known frequency signal. The pattern can either be a circular trace developed in the manner we have described, or it can be a normal representation of the signal displayed by a sawtooth sweep. Examples of each method are shown in Figs. 9-8 and 9-9 respectively.

In the example in Fig. 9-8, the unknown frequency is determined by counting either the light or the dark dashes of the circle (but not both) and multiplying this number by the known frequency used to develop the circular trace. Fig. 9-9 shows a sine waveform marked by the unknown signal. The number of light (or dark) dashes in one cycle of sine waveform multiplified by the known frequency of this waveform will give the unknown frequency. Both these methods work for simple ratios only, that is, ratios where one term is the numeral 1, as in 2 to 1, 5 to 1, 10 to 1, etc. For a ratio like 2 to 7, for example, light spaces will be superimposed upon dark spaces and cannot be interpreted.

ELECTRONIC SWITCH

Two signals can be compared directly on the oscilloscope by first passing them through an electronic switch, as described in Chapter 7. When the two waveforms are superimposed as in Fig. 7-14B, a direct frequency comparison can be made. For example, it can be seen in Fig. 7-14B that five complete cycles of one signal are completed in the time it takes to complete one cycle of the other signal. Therefore, the frequency ratio is 5 to 1. If one frequency is not some exact multiple of the other, one of the waveforms will appear to travel. A slight adjustment of the known frequency will usually correct this effect and allow an easier comparison to be made.

PHASE COMPARISON AND MEASUREMENT

The oscilloscope can be used to make phase comparisons between two signals of the same frequency. The two signals are fed to the vertical and horizontal inputs in the customary manner to develop a Lissajous pattern. The phase relationship between the two signals can be determined by proper interpretation of this pattern.

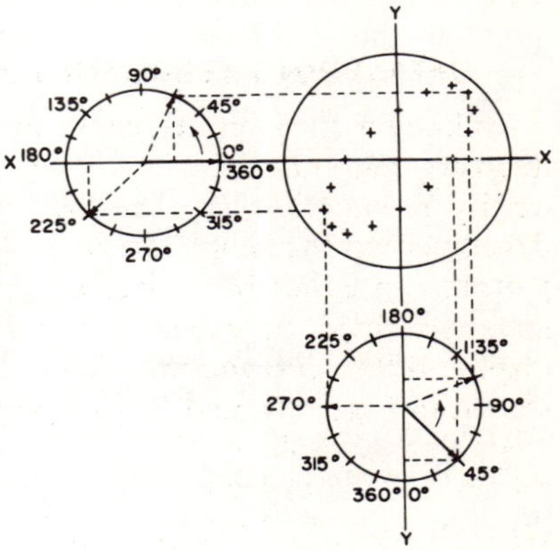

Fig. 9-10. Graphical illustration of the manner in which two sine-wave signals of identical frequency but different phase develop a phase indication on an oscilloscope.

The following discussion and waveforms are based on these arbitrarily chosen phase relationships: (1) the horizontal signal is considered to be leading the vertical signal by the specified amounts, (2) the same number of phase reversals take place in the vertical as in the horizontal amplifiers of the oscilloscope, if amplifiers are used, and (3) a positive-going signal applied to the vertical input causes an upward deflection, and a positive-going signal applied to the horizontal input causes deflection to the right.

Fig. 9-10 illustrates how two sine-wave signals applied to the vertical and horizontal deflection plates develop a pattern indicative of the phase relationship between the two signals. The large circle represents the face of the scope with *X* and *Y* axes drawn through its center. Voltage applied to the horizontal-deflection plates will move the beam along the X-axis to the right or left of center, depending on the polarity of the voltage. Voltages applied to the vertical-deflection plates produce beam movement along the Y-axis.

The two smaller circles represent sine-wave generators for the vertical and horizontal signals and are divided from zero to 360 degrees in steps of 22.5 degrees. The radial arrows are vector representations of the maximum signal voltage of each generator and are made equal to each other for simpler illustration. The deflection factors of the vertical and horizontal plates of the scope are assumed to be equal for the same reason.

In the left-hand circle, a perpendicular to the X-axis from the point of the radial arrow will represent the instantaneous magnitude of the sine-wave voltage applied to the vertical-deflection plates. This value can be transferred graphically to the scope diagram to locate the vertical position of the beam trace at that instant. Other times during the cycle are indicated by dotted arrows. In like manner, the perpendicular to the Y-axis from the arrow point in the lower circle represents the horizontal-deflection voltage at any particular instant. This value is transferred graphically to the scope diagram to locate the horizontal position of the beam trace at that instant.

For this illustration, the two sine-wave vectors have been chosen so that the horizontal vector is 45 degrees ahead of the vertical vector. Since the frequencies of both signals are the same, this 45-degree difference is maintained throughout the complete cycle. The beam position on the scope is plotted for every 22.5-degree interval of the cycle, and the resultant graph gives a very good indication of the scope waveform obtained for a phase difference of 45 degrees between signals. The waveform is an ellipse and is the same as that obtained for a phase difference of 360 degrees minus 45 degrees (315 degrees), as will be seen if we consider the horizontal vector as zero reference, with the vertical vector lagging the horizontal.

Any phase difference can be plotted in this manner, and a complete series would show that the pattern is either a straight line, a circle, or an ellipse. Straight lines occur at zero and 180 degrees phase difference; circles occur at 90 and 270 degrees. All other values of phase difference are shown by ellipses. These ellipses become narrow and approach a straight line at zero or 180 degrees or become broader and approach a circle at 90 and 270 degrees. A number of these patterns are shown in Fig. 9-11. Phase differences are indicated for intervals of 45 degrees, from zero to 360 degrees.

Fig. 9-10 shows how the technician can plot any phase difference he desires and get an accurate waveform like that obtained with the oscilloscope. In practical cases, he may be interested in

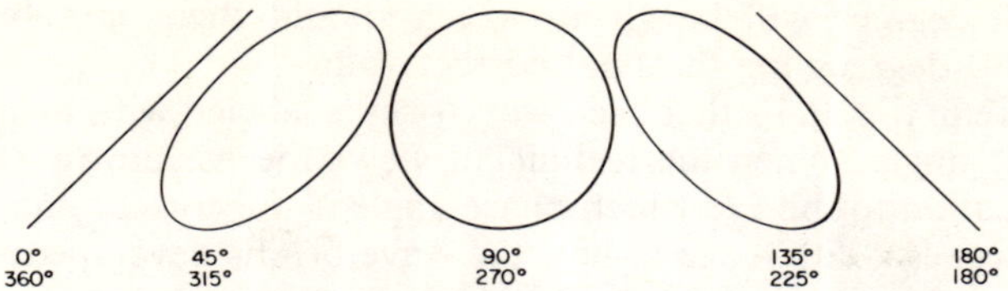

Fig. 9-11. Phase indications from 0° to 360° at intervals of 45°.

somewhat the reverse effect. That is, he may have a Lissajous pattern and desire to know the phase difference it represents. Fig. 9-12 shows how an unknown phase angle can be calculated from such a pattern.

It is convenient to consider all the phase angles from 0 to 90 degrees as basic and to calculate all the others through them. Thus, referring to Fig. 9-11 again, at 0 we start with a straight line (a closed loop seen on edge) and go through an infinite series of ellipses, finally arriving at a circle for 90 degrees. The same series of ellipses, but slanting in the opposite direction, covers the range from 90 to 180 degrees. The range from 180 to 360 degrees duplicates both series, but in reverse order. Before making the measurements of Fig. 9-12, the vertical and horizontal gains should be adjusted to be as nearly equal as possible; otherwise, a perfect circle will not be obtained at 90 degrees. The accuracy of this method will not be impaired, however, if the two gains are not exactly equal.

The ellipse should be positioned so that its center coincides with the intersection of the graph lines of the scope calibration screen as shown in Fig. 9-12. If the distances C and D are measured and substituted in the formula given, we obtain the sine of the phase angle. The phase angle can then be found by locating this value in a table of sines.

Phase angles between 90 and 180 degrees (their ellipses slant from lower right to upper left) are found in the following manner: Find the value of the sine by the preceding method. Locate this value and corresponding angle in the sine table. An angle between

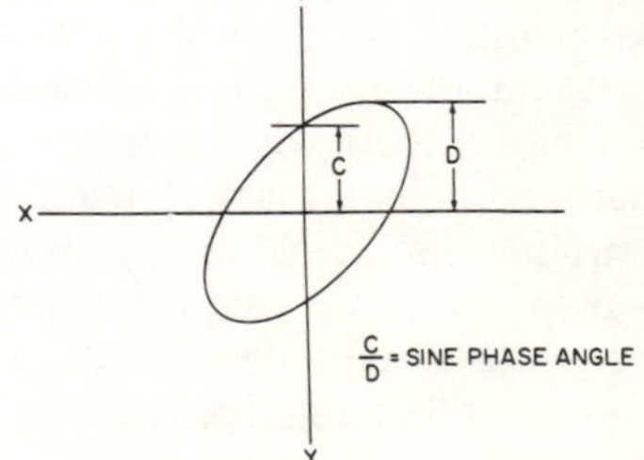

Fig. 9-12. A method for calculating the phase angle represented by a 1-to-1 Lissajous figure.

0 to 90 degrees will be given, which should then be subtracted from 180 degrees for the final correct value.

Notice in Fig. 9-11 that each waveform is labeled with two values of phase angle. When the technician views the waveform, there is no indication to the eye which phase angle is the correct one. However, there is a difference in how the waveform is developed; in one instance, the beam is traveling clockwise around the waveform and in the other, the beam is traveling counterclockwise. Under the conditions stated in the second paragraph of this discussion of phase comparisons, the beam travels counterclockwise for the 45-, 90-, and 135-degree waveforms of Fig. 9-11 and clockwise for the 315-, 270-, and 225-degree waveforms.

The direction of beam travel can be determined by at least two methods: (1) by changing the phase of one signal in a known direction, and (2) by intensity modulation of the beam with a suitable marker. As an example of the first method, suppose we desire to know whether an ellipse like the first, shown in Fig. 9-11, represents a 45- or 315-degree phase difference. We know the horizontal signal is leading the vertical signal by one or the other of these amounts. Suppose we increase the lead of the horizontal signal. If the waveform changes in the direction of the circle, it originally was a 45-degree waveform, but if the waveform changes toward a straight line, it was the 315-degree waveform.

Fig. 9-13 is one example of the type of signal that can be used to mark a pattern so that the direction of beam travel can be de-

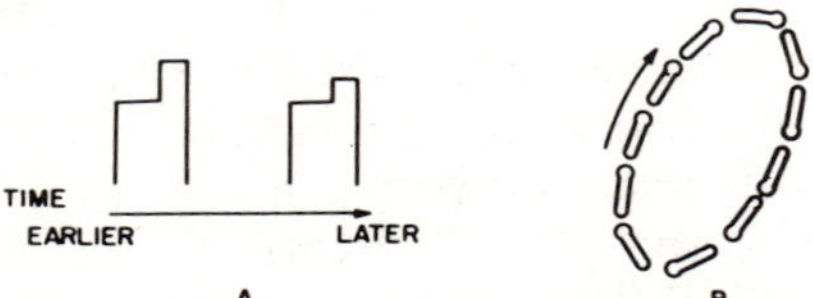

Fig. 9-13. Intensity marker signal at A produces markers as at B. Beam rotation is as indicated by arrow.

termined. This illustration shows just the positive peaks of the signal since they are responsible for the visible indication on the waveform. This type of signal is effective for this purpose because, as it is made more and more positive, the beam trace gets brighter and has a tendency to enlarge. The signal at A in Fig. 9-13 will then result in a marked trace like the one at B. Thus, a trace resembling a series of arrows pointing in the direction of beam travel is obtained. The frequency of the marking signal is unimportant except it should be greater than the vertical input signal and a whole number multiple of this frequency. The intensity adjustment of the oscilloscope should be reduced to nearly minimum intensity for a more effective marker. If a marking signal of the type shown at A in Fig. 9-13 is not available, others may be used. A sawtooth signal like the one developed by the scope sweep generator will produce a narrow wedge-shaped marker. The peak of the sawtooth will correspond to the wide end of the wedge. Which portion of the sawtooth signal occurs later in time must be known to determine the direction of beam travel.

PHASE COMPARISON WITH AN ELECTRONIC SWITCH

Two signals can be displayed together by means of the electronic switch for phase comparisons. They can either be displaced or superimposed for easiest interpretation. Fig. 9-14 shows two examples that are easy to interpret, the 0- and 180-degree phase

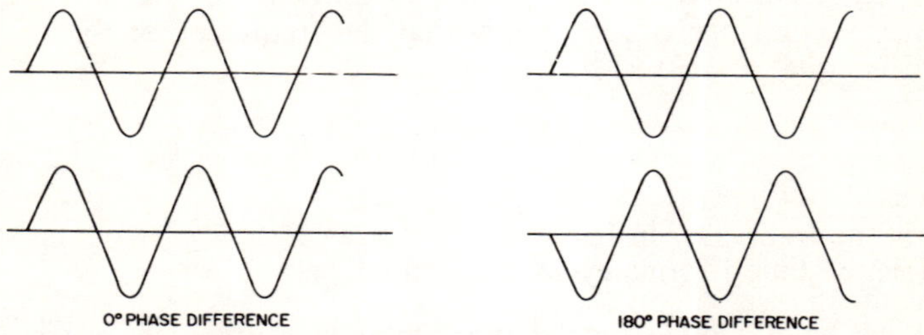

Fig. 9-14. Phase comparisons using an electronic switch.

differences. These figures might be obtained when comparing waveforms at various points in a conventional amplifier. The phase difference between the input and output signals of a single stage of audio amplification usually is 180 degrees.

When viewing small phase differences, the operator may wish to superimpose the two signals as closely as possible. With signals viewed in this manner, the slight phase shift caused by operation of a tone control is large enough to be seen. Both the aforementioned methods of evaluating phase differences, Lissajous patterns and an electronic switch, depend for their accuracy on the skill and care of the operator. At best, the results will be only approximate; the operator will probably do well to measure within 10 degrees of the exact value. However, he can get a good idea of what is happening within a circuit and determine whether or not phase relationships are normal. Special equipment has been designed to give accurate direct indications of phase angles. Some of this equipment involves the use of an oscilloscope, and some does not. Other than this brief mention, however, such equipment will not be discussed here.

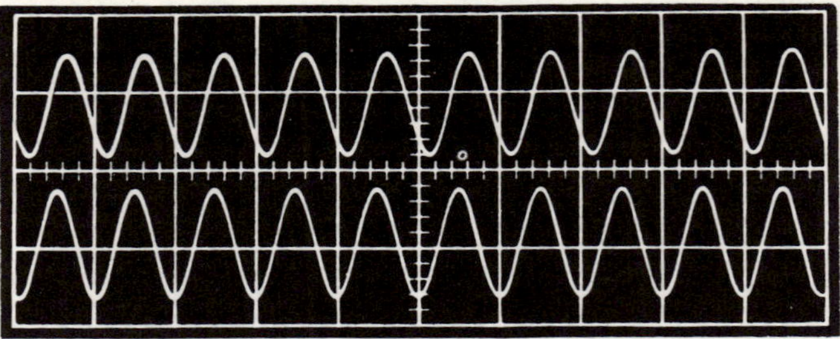

Fig. 9-15. Phase measurement with a dual-trace scope.

Frequency and phase measurements are easily made with suitable lab-type scopes. For example, triggered sweeps are calibrated in time per cm. Thus, if we observe that one cycle of a waveform occupies 4 cm along the horizontal axis when the sweep speed is 1 microsecond per cm, we know that the frequency of the waveform is equal to 250 kHz. In other words:

$$f = \frac{1}{T}$$

where,

f is the frequency in hertz,
T is the time for one cycle in seconds.

If we have a dual-trace scope, phase measurements are easily made as shown in Fig. 9-15. One complete cycle occupies 360°, half a cycle 180°, a quarter cycle 90°, and so on. If one cycle occupies one horizontal division, as in Fig. 9-15, then 1/5 division denotes 72°, or 1/5 of 360°.

Chapter 10

Amplifier Testing with Square Waves and Sweep Signals

A number of amplifier characteristics may interest the technician: frequency response, phase shift, tone control action, stability, equalization pattern, distortion, intermodulation percentage, power output, and others. The first five of these characteristics are particularly suited to exploration by means of square waves and sweep signals. The main advantages of testing with these types of signals are the ease and speed with which it can be done. These testing methods can be applied to audio amplifiers and video amplifiers of monochrome and color-tv receivers. In fact, they should apply to any amplifier whose frequency response falls within the frequency range of the generator and oscilloscope used.

SQUARE-WAVE FREQUENCY RESPONSE TEST

The merit of the square-wave test to indicate an amplifier frequency response is based on the fact that a square wave represents a great many frequencies other than its own fundamental frequency. Fourier analysis has shown that a square wave can be built up from many sine waves of different amplitudes and frequencies. The fundamental frequency will have the greatest amplitude, and it is combined with odd-numbered harmonics that decrease in amplitude as the order of the harmonics increases. Thus, a square-wave signal of a certain fundamental frequency applied to an amplifier is a more extensive test of the amplifier response than is a sine-wave signal of the same frequency. The square-wave test for frequency response consists of applying square-wave signals of various fundamental frequencies to the amplifier and observing the output on an oscilloscope. Any change in the shape of the square wave can then be interpreted in terms of frequency response of the amplifier.

This test is not the type that will furnish information for plotting a response curve. That is, the exact ratio between the amplification factors at various frequencies will not be determined. Rather, a quick overall picture of the frequency response will be gained. Before a test, the quality of the square-wave signal put out by the square-wave generator should be checked. The signal should be fed directly to the oscilloscope input and checked for flatness and sharp corners. A slight imperfection can be tolerated and allowed for when the amplifier response is interpreted. The oscilloscope amplifiers should have response characteristics as good as, or better than, the amplifier being tested; otherwise, they become a limiting factor in the test. If the output waveform from the square-wave generator appears good for some settings of the scope attenuator switch and peaked or rounded at another setting, the attenuator adjustment for that switch position should be checked and readjusted if necessary.

It is customary to assume that a well-reproduced square wave of any particular frequency indicates good response for all frequencies from one-tenth to ten times that frequency. Since this frequency range of 100 to 1 is far short of the range of most present-day amplifiers, the square-wave generator must be reset several times to make a complete test. Usually the extreme low and high ends of the amplifier response will interest the technician more than the midrange. They determine the width of the response range and present more design difficulties to the manufacturer.

An acceptable test is to run the square-wave frequency low enough and high enough until the response falls off at these extremes and then tune the generator through the entire range between, meanwhile watching for any peculiarities.

The technician can acquaint himself with some of the square-wave response curves obtainable if he checks the action of the bass and treble controls during a test. Some of the following waveforms were obtained in this manner. Fig. 10-1 resulted from applying a 1-kHz signal directly to the vertical input of the oscilloscope. The tops and bottoms of the square wave are straight and level, with no evidence of overshoot or ringing. Fig. 10-2 shows the same signal applied to the tuner input of a preamplifier of the type used between tuner or phonograph pickup and amplifier. The bass and treble controls were adjusted for a response approaching Fig. 10-1 as nearly as possible. A slight peak remains at the leading edge of the square wave, and this tendency can be noticed in some of the illustrations which follow. A sine-wave frequency check shows the amplifier to be flat over the audio range. The most likely

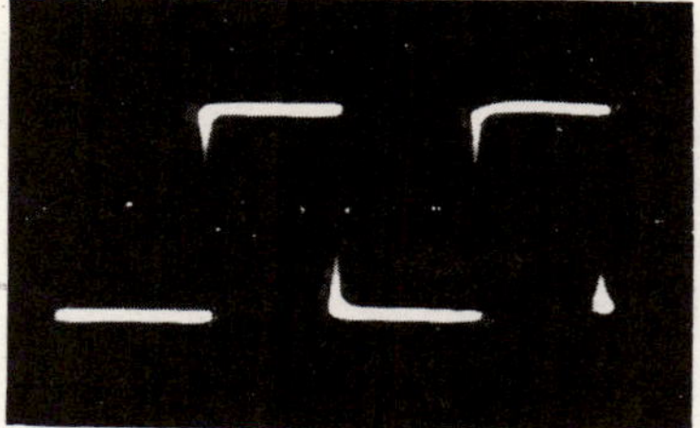

Fig. 10-1. A 1-kilohertz square-wave applied directly to the oscilloscope input.

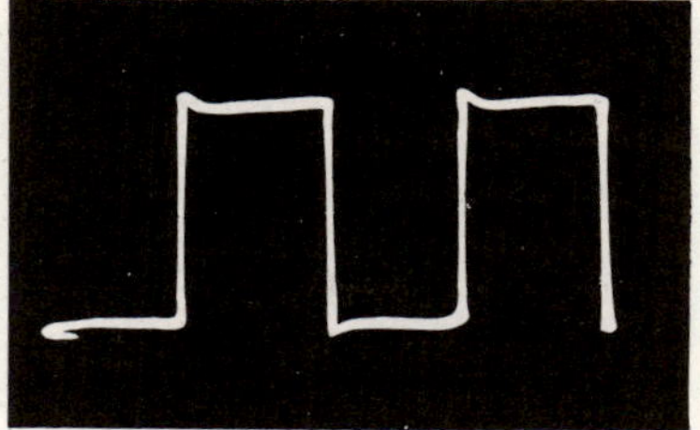

Fig. 10-2. Preamplifier response to the 1-kilohertz square wave. Tone controls were adjusted for flat response.

explanation of the peak is that it is due to some overshoot in the preamplifier circuits.

When the treble control is advanced, the leading edge of each cycle becomes even more peaked, as in Fig. 10-3. The peak appears for both positive and negative halves of the cycle. A logical conclusion from Fig. 10-3 is that excessive highs in an amplifier are indicated by peaked leading edges of the square-wave response, sloping back to the trailing edge. Conversely, the leading edges should be depressed below normal if the highs are attenuated, and this is exactly what happens.

When the treble control is kept normal and the bass control advanced, the waveform of Fig. 10-4 results. Trailing edges of each half cycle of the square wave are peaked, sloping gradually from the leading edges. When the bass control is set for attenuation, the waveform slopes in the opposite direction. It then resembles the waveform for treble emphasis. This is about what you might expect. After all, it is a relative matter—reduce the bass or increase the treble—the results are similar.

Fig. 10-5 is the result of applying a 50-hertz square wave to the preamplifier with the tone controls set for flat response at 1 kHz. Leading edges are elevated and trailing edges are depressed—the marks of low frequency attenuation and phase shift. The severity of a square-wave test over a sine-wave test is borne

Fig. 10-3. Treble emphasis applied to a square-wave signal.

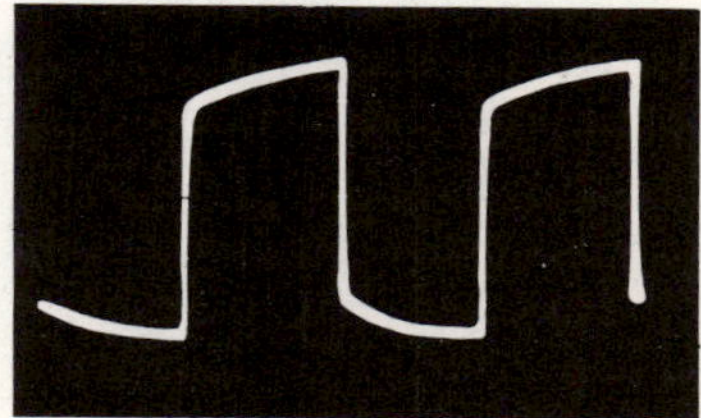

Fig. 10-4. Effect of bass boost on square-wave signal.

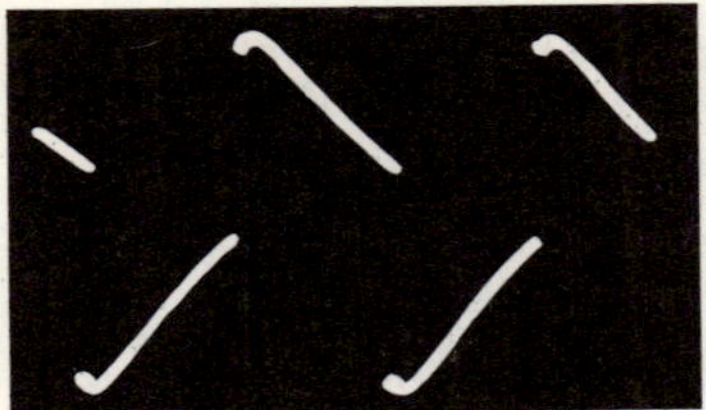

Fig. 10-5. 50-hertz square wave shows low frequency attenuation and phase shift.

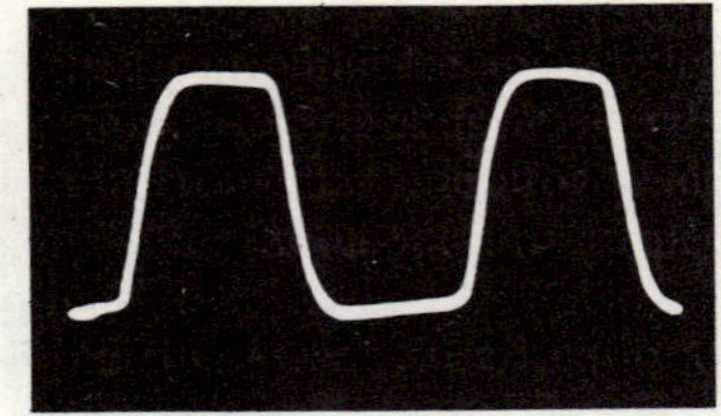

Fig. 10-6. High frequency attenuation is shown by rounded corners of square wave.

out by the fact that the preamplifier showed practically no attenuation to a 50-hertz, sine-wave signal when compared to its 1000-hertz response, yet the 1000-hertz square wave of Fig. 10-1 was changed to that of Fig. 10-5 at 50 hertz.

When the square-wave generator is adjusted toward the high-frequency extreme, a point is reached where the preamplifier response begins to fall. This is evidenced by a gradual rounding of all corners of the square wave, as in Fig. 10-6, which shows the response to a 10-kHz signal. As the square-wave frequency is adjusted higher and higher the corners are rounded more and more, and the square wave takes on the appearance of a sine wave.

THE SQUARE WAVE AS AN INSTABILITY CHECK

We have seen how the square-wave signal provides a quick check for the frequency response range of an amplifier; it can also discover any tendency to instability. The steep wavefront of the square wave can shock borderline cases into ringing or oscillation, as shown by Figs. 10-7 and 10-8. Ringing is just a form of oscillation that dies away quickly. In Fig. 10-7, about 3 cycles of oscillation are visible in each half cycle of square wave. Fig. 10-8 indicates a more unstable condition; the oscillations persist throughout the entire cycle, although they may not show in all parts of the

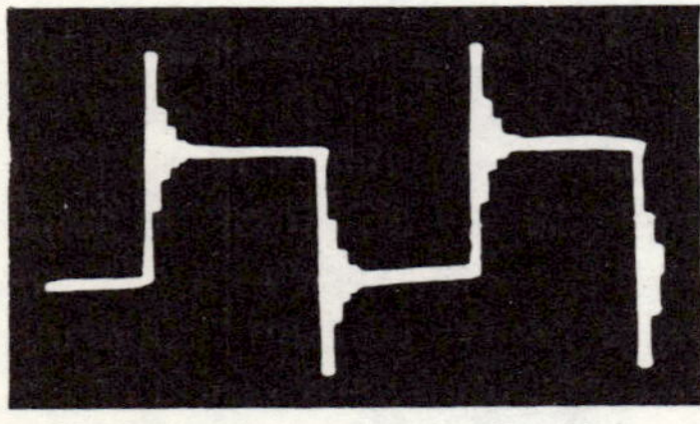

Fig. 10-7. Moderate ringing induced by application of square waves.

Fig. 10-8. Instability shown by continuous oscillation when square wave is applied.

reproduction. When this type of oscillation occurs in audio amplifiers, it is usually above the audio-frequency range and, therefore, will not be heard in itself, but it may react with the audible signal to cause distortion. It is almost certain to cause a lowering of the maximum power output.

Although an amplifier may be shock excited to the point of oscillation with a square-wave signal, it might not do so when a sine-wave signal is applied. The results of the square-wave test indicate, however, that some trouble might be expected on large audio signals of complex waveforms.

Video amplifiers in tv receivers can be tested with a square-wave signal in much the same manner as audio amplifiers. Interpretation of the resultant waveforms is similar to the examples just given. The range of frequencies is a little different with the tv receiver; it is shifted toward the higher frequencies. That is, it is usually not expected to go as low, but it does extend up to several megahertz. Here again, the operator must be certain that the oscilloscope amplifiers do not become the limiting factor in response at these higher frequencies, or a misleading waveform will be obtained.

Phase shift and ringing are more evident in the tv picture than in an audio signal and cause smearing and repeated outlines during reception of a regular broadcast signal. Incidentally, the picture tube can serve as a fair substitute for the scope in the square-wave test since the signal applied to the video amplifiers is fed directly to it. By turning the brightness up and down, the operator can get a good idea of the condition of the square-wave signal at the picture tube. For example, sharp corners on the waveform will result in sharp divisional edges between the light and dark bars on the screen; round corners will result in blended edges to the bars.

TESTING WITH SWEEP SIGNALS

The sweep method of testing results in waveforms that approach the conventional plotted response curve of an amplifier—at least in content if not in actual appearance. The test works for both audio and video frequency amplifiers, if a suitable sweep signal is available. Radio-frequency sweep generators have been common for many years now, but video and audio sweep generators are somewhat less common. One method of designing sweep generators, beating a steady signal against a sweeping signal, is difficult to use at low sweep frequencies. The reason is that two oscillators tend to lock together when they approach each other in frequency. Thus, sweeps at comparatively low frequencies, such as the audio and video frequencies, are a little more difficult to attain.

One method of obtaining an audio sweep signal is with a sweep recording. The recording is in the conventional disc form and must be played on a record player to obtain the signal. When the playback system has the proper equalization characteristics, corresponding to the recording characteristics originally used in making the recording, a flat output signal is obtained. The operator can check pickup characteristics, preamplifier equalization, and tone control action. To make the signal more useful, markers are included during the original recording process. These markers identify the frequencies at various points on the response curve. Fig. 10-9

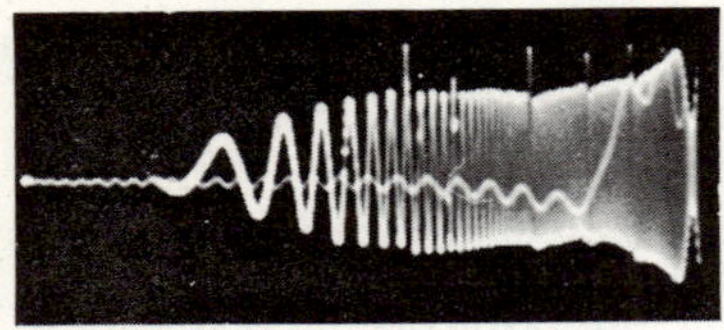

Fig. 10-9. Sweep signal obtained directly from disc recording with variable reluctance pickup.

shows a response curve obtained by playing an audio sweep recording with a variable reluctance cartridge. No equalization or amplification was used between the cartridge and the oscilloscope. The recording was cut to a modified NARTB curve, with markers at 70, 1000, 3000, 5000, 7000, and 10,000 hertz.

Since the cartridge was not equalized, the response waveform is not flat, but shows a reduced response at the low end and a peaked response at the high end. The cartridge is velocity responsive, and would produce a flat waveform if the recording had been made at constant velocity. However, the recording has a constant-amplitude section at the low end and a section of preemphasis at the high end, and these sections account for the deviations in response.

The oscilloscope was set for internal sawtooth sweep of 20 Hz, with internal synchronization. The 70-hertz marker does not appear in the waveform and has probably been lost at the beginning or end of the oscilloscope sweep. This type of waveform is a little difficult to synchronize so that the whole waveform appears in order from low to high frequencies, as it does in this photo. More often than not, the scope sweep will start somewhere near the middle of the response, and all of the waveform which should appear to the left of this point will be displaced to the right-hand side of the waveform. This is just a minor inconvenience, since the waveform can still be interpreted without much trouble.

The audio sweep record provides a quick and convenient method of checking the effectiveness of equalization networks and tone controls. One thing should be remembered—the playback cartridge is necessary to develop the signal, and since its effect on the signal has a bearing on the final waveform, it must always be considered as one link in the amplifier chain. Other methods of sweep generation do not require a playback cartridge, and so this factor is eliminated in the final consideration of the waveform.

CHECKING VIDEO RESPONSE

A complaint frequently encountered by service technicians is that a television receiver produces a picture lacking in fine detail. Since the finely detailed portions of the picture are produced by the higher video frequencies, this lack of fine detail indicates that these high frequencies are not being presented to the picture tube. If the r-f amplifier, mixer, and i-f stages have been checked, and are operating satisfactorily, the trouble must lie in the video amplifier. The video amplifier can be checked with a voltmeter, but this will not prove conclusively that the operation is normal.

The requirements imposed on a video amplifier are quite strict. It must amplify equally all frequencies from 30 hertz to over 4 MHz and still maintain an average gain of 20 to 30. That is, the amplifier must have a flat frequency response and sufficient gain. It is difficult to produce both, for if one is increased, the other will decrease, and vice versa.

Any variation in component values may change the frequency response of the amplifier, and to detect this change, the frequency response must be known. A graphic curve can be constructed from values obtained by measuring the amplification at various fixed frequencies, but this can become a laborious and time-consuming process. A much simpler and easier method of determining the video-amplifier response is to employ an oscilloscope and sweep generator, in much the same manner as is done in video i-f alignment.

The output of an fm signal generator is a frequency-modulated signal that varies between upper and lower frequency limits determined by the setting of the controls. The setting of the center-frequency control determines the frequency around which the signal deviation occurs. The sweep-width control is used to set the amount of desired deviation above and below the center frequency. For instance, a setting of 25 MHz on the center-frequency dial and a 10-MHz sweep-width setting provide a frequency-modulated signal between 20 and 30 MHz. When used to align a tuned amplifier, such as the video i-f stages, the center frequency of the generator is adjusted to the center of the amplifier passband, and the sweep limits are adjusted to cover the upper and lower limits of the passband. The amplifier output can then be displayed on an oscilloscope screen in the form of a curve. The synchronized sweep voltage from the generator must be connected to the horizontal-input terminals of the oscilloscope. The complete horizontal trace then represents the frequency band covered by the sweep generator. Any point on the horizontal trace represents a definite frequency; therefore the height of the curve at any given point represents the amplifier output at that frequency. Thus, a curve representative of the amplifier gain at all frequencies within the passband can be displayed. This method can also be used for testing video amplifiers by adjusting the generator to sweep from zero cycles up to the maximum frequency to be checked.

Many combinations of generator and oscilloscope will provide a usable pattern, and others will not. At this point, the equipment on hand should be checked to be sure the generator has enough output and the oscilloscope has a sufficiently wide response. The generator and oscilloscope are connected and the controls adjusted as if an amplifier were to be tested, except that the generator output is connected directly to the oscilloscope input. This gives a response check of both units and should produce a pattern similar to that of Fig. 10-10. This photo shows the response curve pro-

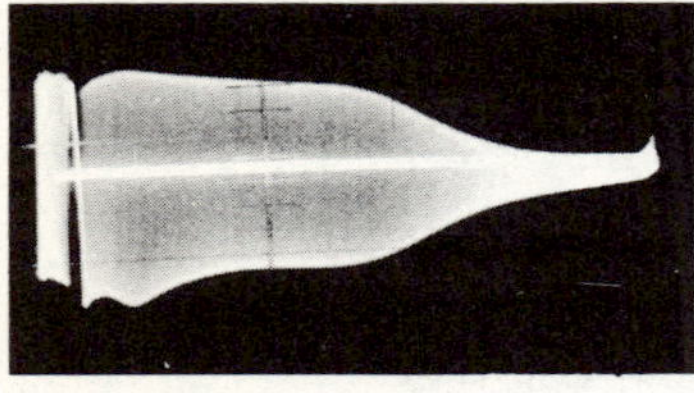

(A) Wide-band oscilloscope.

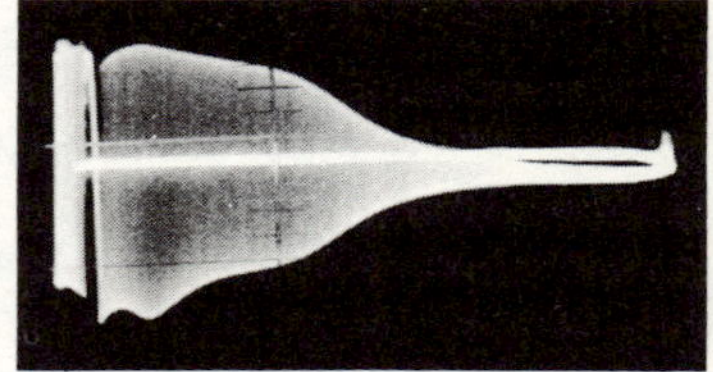

(B) Narrow-band oscilloscope.

Fig. 10-10. Response curve with generator directly connected.

duced by the generator and oscilloscope, which were used to procure all the included photographs. Fig. 10-10 shows a curve obtained with an oscilloscope having insufficient high-frequency response. If a check of the equipment on hand produces a similar pattern, the oscilloscope is not suitable for checking video-amplifier response in this manner.

Connect the sweep generator to the amplifier input, and adjust the controls to provide an output of a 4.5-MHz center frequency with a sweep width of 9 MHz. This setup was used to obtain the accompanying photographs. A wide-band oscilloscope is connected to the amplifier output, and the synchronized sweep output of the generator is connected to the horizontal-input terminals of the oscilloscope. The oscilloscope controls are adjusted so that the sweep voltage from the generator provides the horizontal trace. With this setup the oscilloscope screen presents a response curve of the

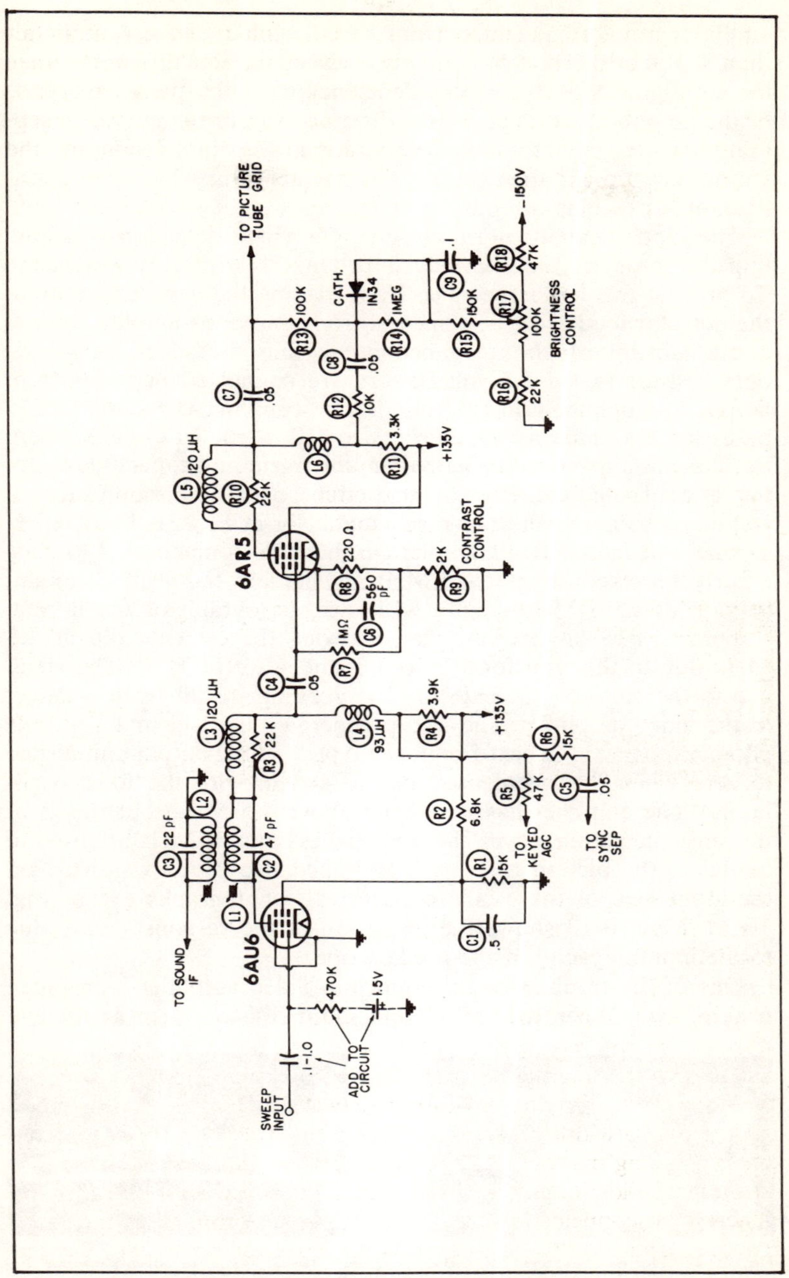

Fig. 10-11. Schematic of video amplifier used in test.

amplifier at all frequencies, from a few kilohertz to a little better than 4.5 megahertz (the generator sweep is actually wider than the nominal 9 MHz), and any deficiency in the frequency response of the amplifier will appear as a droop or sag in the curve. A separate marker generator may be coupled to the amplifier input, and the marker pip will then identify the frequency at which the amplification loss begins or ends.

The diode detector normally used for video detection is a low-impedance device, and its load impedance is of a very low value. To prevent this low impedance from loading the generator output, the detector load must be removed from the video-amplifier input, and a substitute high-impedance circuit must be added. A 470K-ohm resistor and a 1.5-volt battery are connected in series from the video-amplifier grid to ground. The coupling capacitor should be as large a value as possible (.1 to 1.0 μF). In circuits where fixed bias is applied to the video-amplifier grid and capacitive coupling is used, only the detector-load circuit need be disconnected.

The video amplifier shown schematically in Fig. 10-11 is representative of most video amplifiers in that it is compensated to have a fairly flat response up to 4.5 MHz and actually has a small amount of gain up to 6 MHz. Fig. 10-12 is a photograph of the normal response curve of this amplifier, showing the extreme dip at 4.5 MHz due to the trap formed by L1 and C2. (NOTE: The small gap in the curve at the extreme left does not result from a defect in the video amplifier. The sweep generator used in making these photographs was the beat-frequency type, and the output diminished to zero when the swept oscillator locked in with the fixed oscillator.) The amplifier has some gain above 4.5 MHz, but this is of no consequence since only the frequencies below 4 MHz are used to modulate the picture tube. A high-impedance probe was used on the input lead of the oscilloscope for all photographs except Fig. 10-13. Fig. 10-13 shows the distortion of the response curve due to the input capacity of the oscilloscope.

One of the troubles often found in a video amplifier is an open peaking coil. When the coil has no shunt resistor, such as L4 and

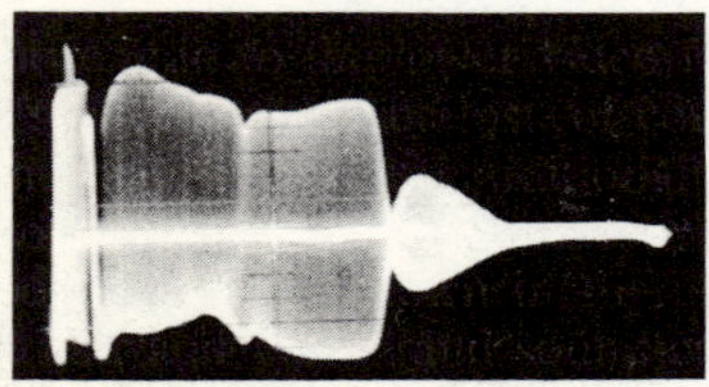

Fig. 10-12. Normal response of video amplifier.

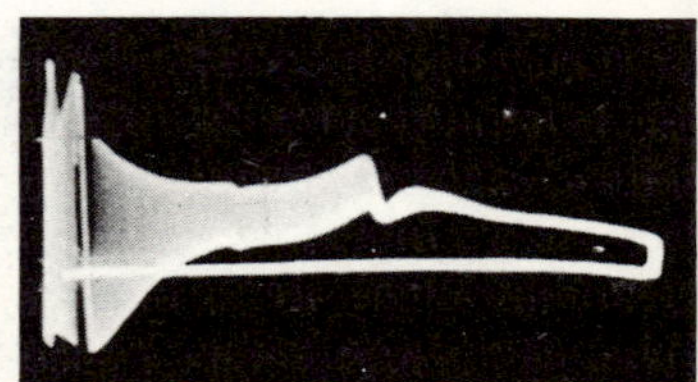

Fig. 10-13. Response curve showing distortion introduced by input capacity of oscilloscope.

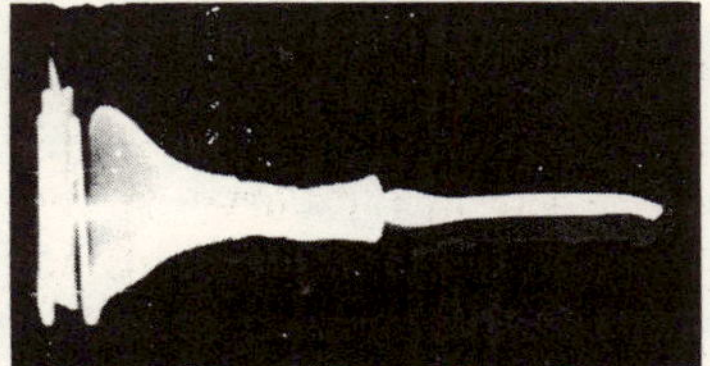

Fig. 10-14. Response curve produced by open in series-peaking coil in 6AU6 plate circuit.

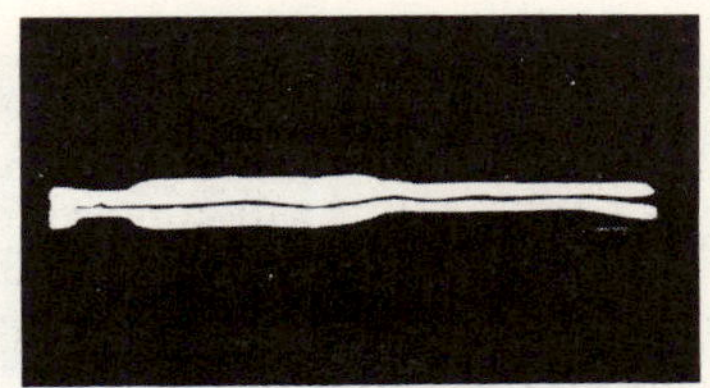

Fig. 10-15. Response curve produced by open in series-peaking coil in 6AR5 plate circuit.

L6 in Fig. 10-11, the result of an open will be definite. The plate voltage is removed from the tube, and the amplification ceases. The result of an open in L3 or L5 of Fig. 10-11 will be less definite. The shunt resistor is still in the plate circuit; the tube retains plate voltage, although at a low value, and some amplification remains. Fig. 10-14 shows the result of an open in L3, and Fig. 10-15 shows the result of one in L5. Capacitor C6 in the cathode circuit of the 6AR5 video-output stage, was included in the amplifier design to improve the high-frequency response. If an open should occur in this capacitor, the oscilloscope pattern would be similar to Fig. 10-12, but would have a lower amplitude. However, should this capacitor develop an internal short, it would remove the normal bias from the 6AR5, in addition to lowering the response. Fig. 10-16 illustrates the result of this condition, showing the distortion caused by operating the 6AR5 at approximately zero bias.

While a receiver is being serviced, leads and components may have to be rearranged. A lead or component might be moved so close to the video amplifier circuits that an appreciable capacity to ground would be introduced into the circuit. The plate circuits of the video amplifier would be most affected by this added stray capacity. To simulate this condition, a 5-pf capacitor was shunted to ground from two separate points in the video amplifier, and the resulting waveforms appear in Figs. 10-17 and 10-18. Fig. 10-17 was obtained with the capacitor connected to the plate (pin 5) of the 6AU6, whereas in Fig. 10-18, the capacitor was connected to

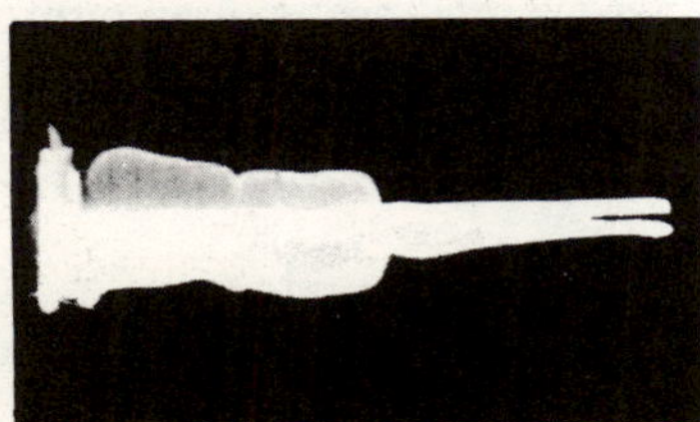

Fig. 10-16. Response curve with cathode bypass capacitor shorted.

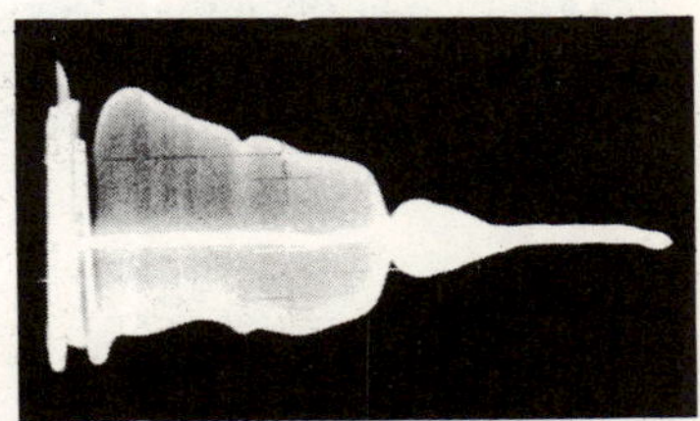

Fig. 10-17. Response curve showing effect of added capacity.

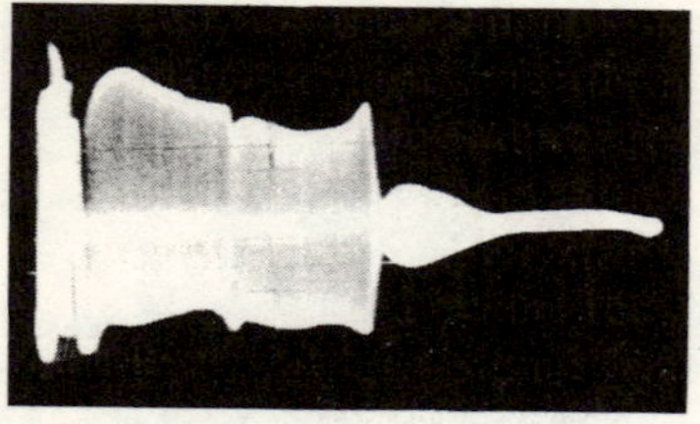

Fig. 10-18. Effect of stray capacity on the high-frequency response of an amplifier.

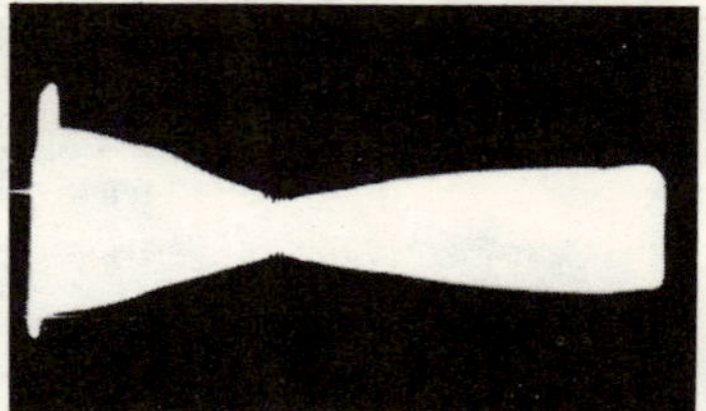

Fig. 10-19. Response curve obtained with correct adjustment of 4.5-megahertz trap.

the junction of L1 and L3. The added capacity distinctly affects the high-frequency response of the amplifier. This condition could have previously been injected into a receiver by a service technician who had thoughtlessly moved a lead or component.

This method of checking video response also provides a fast and accurate way of adjusting or checking the 4.5-MHz trap during the response check. Reduce the Sweep-Width control setting to approximately 1 MHz; adjust the Center-Frequency control to center the response curve on the oscilloscope screen, and inject a calibrated 4.5-MHz signal into the video amplifier input. Fig. 10-19

Fig. 10-20. Response curve showing misadjustment of 4.5-megahertz trap.

shows the pattern obtained when the trap is correctly adjusted, and Fig. 10-20 shows an incorrect adjustment.

Although this procedure may seem complicated at first, with repeated usage the complications should disappear. It will be easier to set up the equipment and obtain an overall indication of amplifier performance at once than to make numerous readings with an ohmmeter or voltmeter. For the service technician who demands the best performance from the receivers he has serviced, this method should prove valuable.

Chapter 11

Radio and TV Alignment

There are a number of tunable circuits in the average radio or tv receiver. Their purposes vary, but usually they are designed to accept or reject certain frequencies so they can be amplified or eliminated entirely, or otherwise controlled. The adjustment of these circuits for proper functioning is called alignment. Several choices may be open to the technician for methods of alignment, but the choice of indicating equipment will usually be between meter and oscilloscope. Sometimes a combination of both may work better. Since this book deals primarily with the oscilloscope and its uses, alignment discussion will be restricted mainly to this type.

ADVANTAGES OF OSCILLOSCOPE ALIGNMENT

Although a simple a-m radio receiver can usually be satisfactorily aligned with a meter, the advantages provided by an oscilloscope are readily apparent when more complex and exacting alignments are attempted. Examples include wide-band i-f amplifiers of both am and fm radios, fm detectors, stagger-tuned or overcoupled video i-f amplifiers, and *Synchroguide* circuits. Alignment of a wide-band i-f amplifier with a meter would involve many individual settings of an rf generator and checks of the effect after each alignment adjustment. With a scope and sweep generator, a complete response curve is seen, and the effect of any adjustment upon the entire response is noticed immediately. In addition, overloaded circuits result in distortion, which cannot be seen on a meter but can be seen on a scope. Also, meters are more susceptible than oscilloscopes to damage from overloads.

PRELIMINARY STEPS

It is good practice to use an isolation transformer between the power line and any ac/dc receiver being aligned to protect both operator and test equipment. The ground lead of much test equipment is connected directly to the case. If the ground lead is connected to any live point on an ac/dc chassis, the results may be unpleasant, to say the least, since it is difficult to avoid touching the case during alignment.

If the receiver being aligned has an avc or agc circuit, the alignment instructions will usually recommend disabling or controlling this circuit during alignment. One reason for this is that the avc or agc action will partly nullify the effect of an alignment adjustment. Adjustments are usually made for a maximum or a minimum output indication at the scope, but the agc tries to hold the output constant. Therefore, the agc should either be disabled or should be held at a constant value.

When a circuit is fairly simple and has no bandwidth considerations, one of the most satisfactory procedures is to maintain the input signal from the alignment generator as low in level as possible and thus avoid any action from the avc circuit. This system can also be followed where a bandpass effect, together with maximum sensitivity, is desired, for example, when a tv receiver is aligned for fringe area reception.

When the receiver is aligned for reception of normal strength tv signals, rf and video i-f stages should be maintained at the same bias level that would be obtained with signals of this strength. In some amplifiers, the overall response may change as the operating bias is changed. Therefore, to ensure best response during actual reception, the receiver is aligned with normal operating bias. This also helps avoid adjustment of any stage to an overload condition.

The alignment bias is usually applied to some convenient point on the agc line and can be a small battery or other bias supply. Some supplies are made to be adjustable so that a range of values can be had. The correct value and point of application are usually suggested by the receiver manufacturer. Sometimes the value is not considered critical and is specified as "that value of bias which results in an undistorted waveform." Fig. 11-1 shows part of a tv receiver schematic that can serve as the basis for discussion in the following paragraphs. The section which carries the signal from the antenna straight through to the video output stage is shown and includes the tuner and video i-f amplifiers. The alignment instructions for the video i-f section appear in Fig. 11-2 and the desired response curve in Fig. 11-3. The recommended bias for the tuner is 1.5 volts, applied to point ⟨A⟩ . The first and second video i-f

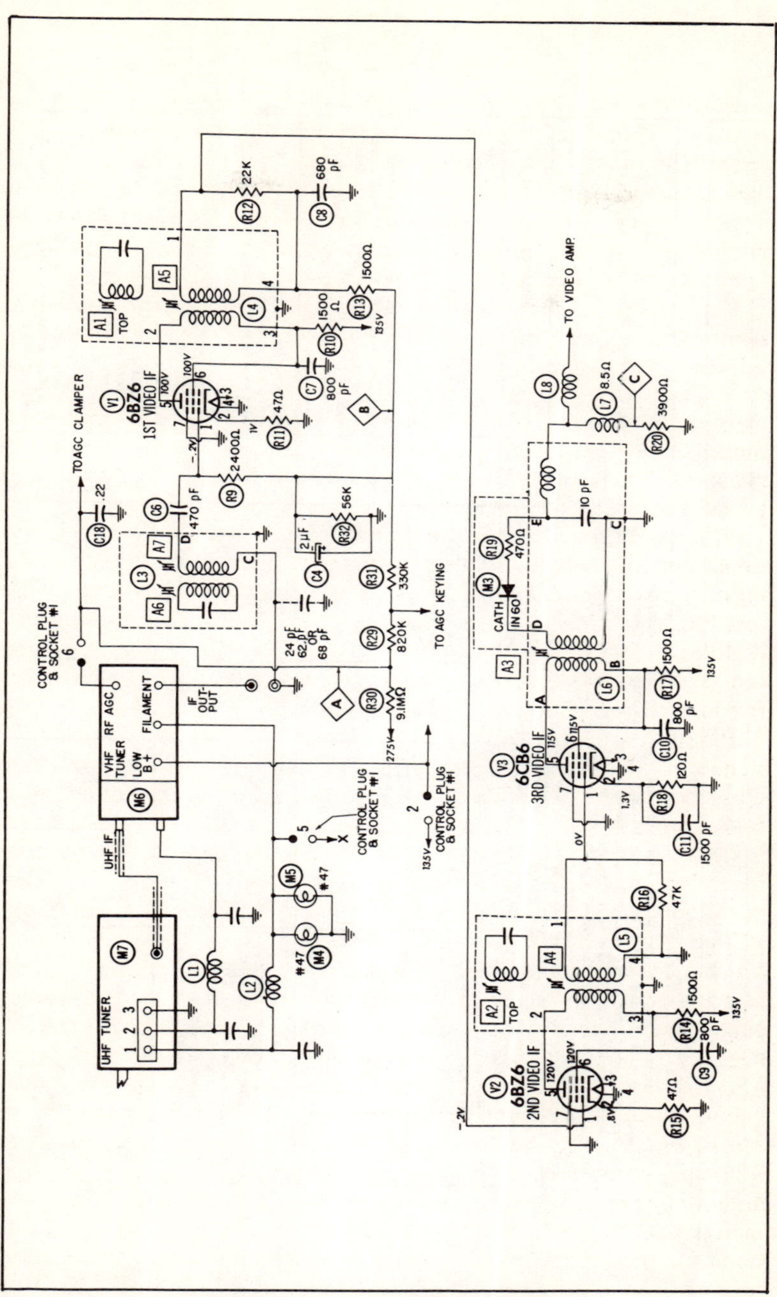

Fig. 11-1. Video i-f channel of a tv receiver.

PRE-ALIGNMENT INSTRUCTIONS

The high-voltage lead should be securely taped and kept away from the chassis.
Allow a 20-minute warm-up period for the receiver and test equipment.

VIDEO IF ALIGNMENT

Connect the negative lead of a 1.5-volt bias supply to point Ⓐ. Positive to chassis.
Connect the negative lead of a 3-volt bias supply to point Ⓑ. Positive to chassis.
Connect the synchronized sweep voltage from the sweep generator to the horizontal input of the oscilloscope for horizontal deflection.
The sweep generator output lead should be terminated with its characteristic impedance, usually 50 ohms.
Use only enough sweep-generator output to provide a usable pattern on scope.
Use 10-mc sweep unless otherwise stated.
Detune mixer plate coil by turning core fully counterclockwise.

	DUMMY ANTENNA	SWEEP GENERATOR COUPLING	SWEEP GENERATOR FREQUENCY	MARKER GENERATOR FREQUENCY	CHANNEL	CONNECT SCOPE	ADJUST	REMARKS
1.	.001 μF	High side to pin 1 (grid) of 6BZ6 (V1). Low side to chassis.	43.5 MHz	47.25 MHz	Any non-interfering channel	Vert. Amp. thru 10K to point Ⓒ. Low side to chassis. (Across Video Det. load)	A1, A2	Adjust to place marker in trap notch. If two points are found to do this, use the one with slug farthest counterclockwise.
2.	"	"	"	42.25 MHz	"	"	A3, A4, A5	Adjust for maximum gain and symmetry of response similar to Fig. 11-2B with markers as shown. Adjust A3 for maximum gain, A4 to position 45.75-MHz marker and A5 to place 42.25-MHz marker. Recheck step 1.
3.	Direct	Place a thin insulated metal strip between the Mixer-Osc. tube (V202) and tube shield. Connect the high side of sweep generator to the metal strip. Low side to chassis.	"	41.25 MHz	"	"	A6	Adjust to place marker in trap notch. If two points are found to do this, use one with slug farthest counterclockwise.
4.	"	"	"	41.25 MHz 42.25 MHz 45.0 MHz 45.75 MHz	"	"	A7 & Mixer Plate Coil	Adjust for maximum gain and symmetry of response similar to Fig. 11-2B with markers as shown. Adjust mixer plate coil for maximum gain with 45.74-MHz marker at 50%. Adjust A7 for maximum gain and proper tilt. Due to interaction, it may be necessary to repeat adjustment. Recheck step 3.

Fig. 11-2. Alignment instructions for receiver circuit of Fig. 11-1.

stages are biased at 3 volts, applied to point ◇ . Resistors R29, R30, and R31 isolate these bias supplies from each other and from the 275-volt source.

Another preliminary step often recommended before performing a video i-f alignment or response check is to disable the local oscillator in the tuner section. Converter action is not needed for the alignment or response check, and the oscillator, if allowed to operate, may beat with the video i-f signal supplied by the generator. The result can be a number of extra curves that resemble the desired response curve and add to the general confusion. A recommended method of disabling the local oscillator is to sub-

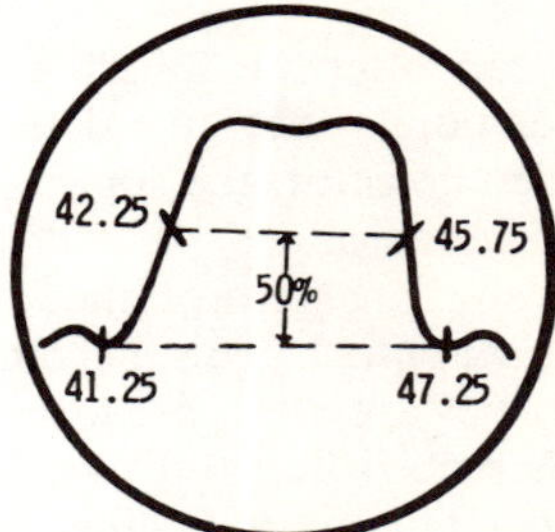

Fig. 11-3. Video i-f response curve.

stitute a mixer tube of the same type, but with the oscillator grid or plate pin clipped. This method is a practical one for the technician who handles a large number of alignments. A small number of tubes so treated will serve for a large percentage of receivers aligned. If, for any reason, disabling the local oscillator does not seem practical, it can be left operating and the channel setting and fine tuning can be adjusted for minimum effect on the response curve. Avoid channels occupied by local television stations.

CONNECTION POINTS AND METHODS

Preparation for an oscilloscope alignment includes the connection of sweep generator and oscilloscope to the circuits. Fig. 11-4 shows the minimum equipment necessary to obtain the response curve of any circuit in a receiver. The rf sweep signal is applied to some point of the receiver ahead of the circuit to be analyzed, and the oscilloscope vertical input is connected to some point following it. The resultant waveform indicates the response to the applied sweep frequencies of all circuitry between these two points. A detector is shown between receiver and oscilloscope in Fig. 11-4. This detector is necessary when the sweep frequencies are higher than the response characteristics of the scope amplifiers. Such a de-

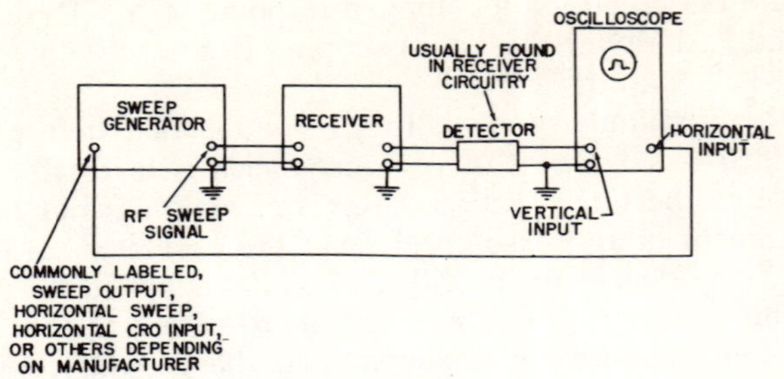

Fig. 11-4. Minimum equipment necessary to obtain a response curve.

tector can often be found at some point in the receiver. Typical receiver points that give detection are the mixer grid of the tuner, the video detector, the sound i-f limiter grids, and the ratio detector or discriminator.

The sweep generator should be connected so that an undistorted signal is applied. This becomes more important as higher sweep frequencies are used, because the connecting cable has transmission line characteristics and must be correctly terminated to avoid reflections. This point is illustrated in Fig. 11-5A, which shows the response of the rf stage of a television tuner. The signal input was made to the receiver antenna terminals through the output cable supplied with the sweep generator. This cable had a built-in terminating network, which is diagramed in Fig. 11-5C. The generator sweep was set for Channel 4, with a 12-MHz sweep and a 68-MHz

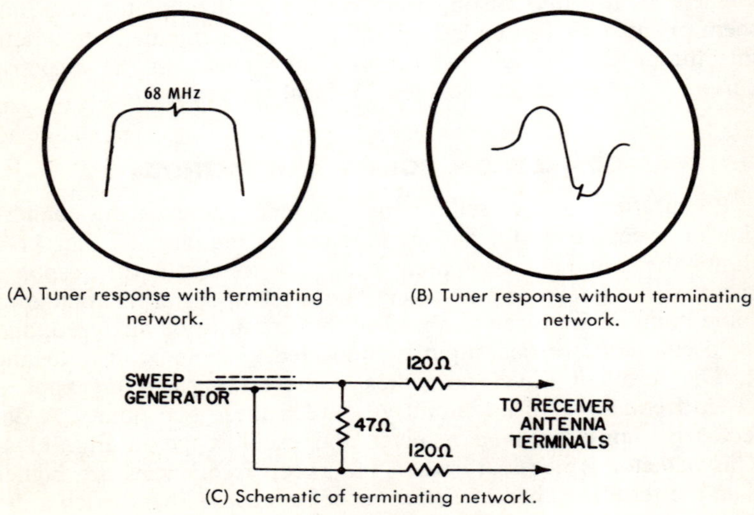

(A) Tuner response with terminating network.

(B) Tuner response without terminating network.

(C) Schematic of terminating network.

Fig. 11-5. Matching network and response.

marker. The network shown matched the characteristic impedance of the output cable to the 300-ohm impedance of the antenna input terminals. When a shielded output cable with no terminating network was substituted for the regular cable, the distorted response shown in Fig. 11-5B was obtained.

Examples of a few sweep attenuator pads recommended for use during alignment are shown in Fig. 11-6. They are designed to match three different generator-cable impedances to the 300-ohm balanced input of tv receivers.

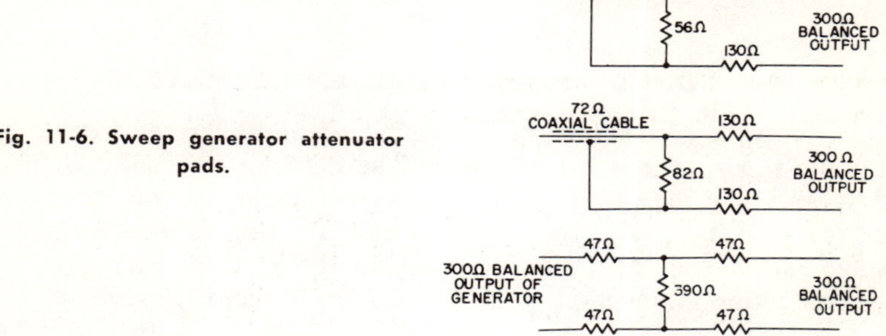

Fig. 11-6. Sweep generator attenuator pads.

Sweep generator connection to the receiver for tuner alignment is usually made directly to the antenna terminals through a matching network like those shown in Fig. 11-6. Connection points and methods can be somewhat different for video i-f alignment. A common point at which to introduce the video i-f signal is at the mixer tube. The signal then passes through any tuned circuit in the plate circuit of this tube, and the response of the entire video i-f section can be viewed at the video detector. There are several ways to inject the sweep signal into the mixer tube circuit; the sweep generator lead can be clipped directly to the mixer grid terminal (not to the oscillator grid), to an ungrounded tube shield over the mixer tube, or to an insulated metal strip inserted between tube shield and tube. The third method has the advantage of the tube not being left unshielded. The metal strip can be moved about to find the most sensitive position. Some tv tuners have a lead brought out to the top surface of the tuner from the mixer grid as an alignment convenience. This point is commonly called a "looker" point and can be used either to view the rf response curve or to inject the i-f sweep signal.

The matching networks of Fig. 11-6 are not used when the video i-f signal is injected at the mixer stage or succeeding stages. The sweep generator cable should be terminated with a resistance equal to its characteristic impedance, usually 50 or 75 ohms, to avoid cable reflections.

Sometimes a video i-f alignment is performed one stage or more at a time, instead of as a group. For example, if the sweep signal is applied to the grid of one i-f stage and the scope is connected to the grid of the next stage, the response curve of the first stage can be obtained. A detector circuit is used between the scope input and the video i-f stage. Tuned circuits in preceding and succeeding stages may affect the response curve; therefore, some manufacturers recommend shunting these tuned circuits with a low-value resistor to reduce any effect on adjacent circuits during alignment. Values of 180 to 330 ohms are commonly recommended.

FM SOUND I-F AMPLIFIERS AND DETECTORS

A scope alignment of the fm sound i-f amplifiers and detector usually separates the two stepwise. The detector is usually aligned first, and then the i-f amplifiers are aligned; however, the order can be reversed if some of the i-f stages have limiting action. A limiter stage can be used as a viewing point for the i-f response curve.

The scope connection point for an fm detector alignment depends on what type of detector it is—that is, whether it is a ratio detector or a discriminator. Fig. 11-7 shows the sound i-f circuit for a tv receiver and includes a typical ratio detector circuit. The scope is connected at point ⟨E⟩ while the primary of the ratio detector transformer is being adjusted, and at point ⟨F⟩ while the secondary is being adjusted. Stabilizing capacitor C3 is 5 μF, a value large enough to bypass some of the response curve (the generator sweep rate is usually 60 Hz). Therefore it must be disconnected while the scope is at point ⟨E⟩ . For this particular circuit, it is convenient to adjust A12, A13, and A14 at the same time as A15, although separate adjustment could be made by placing the scope across R57 (limiter grid resistor).

The sweep generator can be connected to point ⟨B⟩ in Fig. 11-7 if it is desired to feed the signal through the sound i-f stages, or it can be connected to point ⟨D⟩ if the ratio detector is to be aligned separately.

A discriminator circuit is shown in Fig. 11-8. Oscilloscope alignment of this fm detector can be accomplished with the scope connected to one point, ⟨C⟩ , while slugs A11 and A12 are adjusted.

MARKERS

An amplifier response curve displayed on an oscilloscope screen may cover a wide range of frequencies. For maximum usefulness,

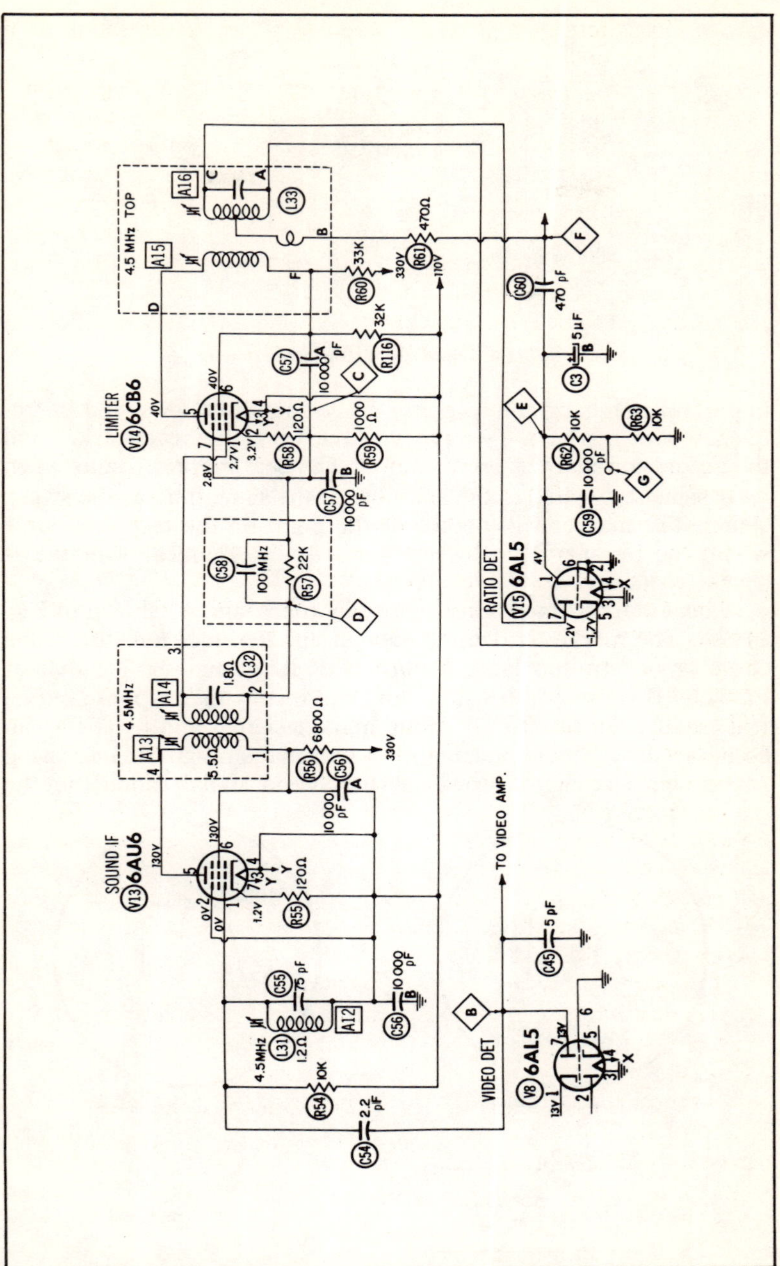

Fig. 11-7. An fm sound i-f channel and ratio detector.

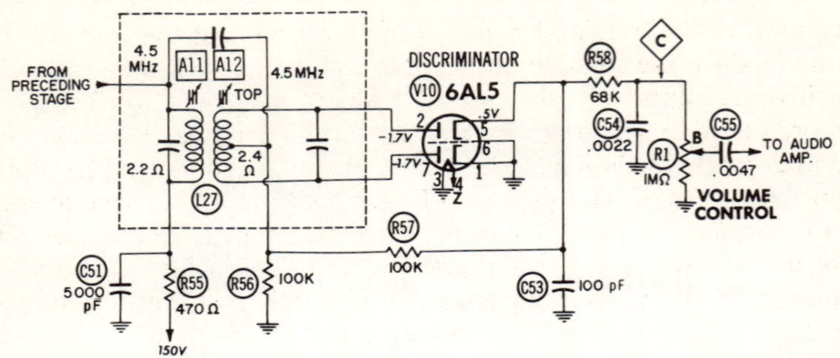

Fig. 11-8. Typical fm discriminator circuit.

important points on the response curve should be identified in frequency with markers. Two types of markers, the beat marker and the absorption marker, are common. The beat marker results when an rf signal is applied to the amplifier at the same time as the sweep signal. The marker will appear at the point on the response curve where the instantaneous sweep frequency is identical with the rf signal frequency.

A beat marker that is satisfactory in appearance is shown in Fig. 11-9A. The marker is definite enough and yet does not distort the curve or occupy too large a portion of it. A marker like that of Fig. 11-9B is undesirably large and can distort the response curve. It is usually the result of a strong marker signal, and its effect can be reduced by either reducing marker signal strength or increasing sweep signal strength. However, the sweep signal should not be

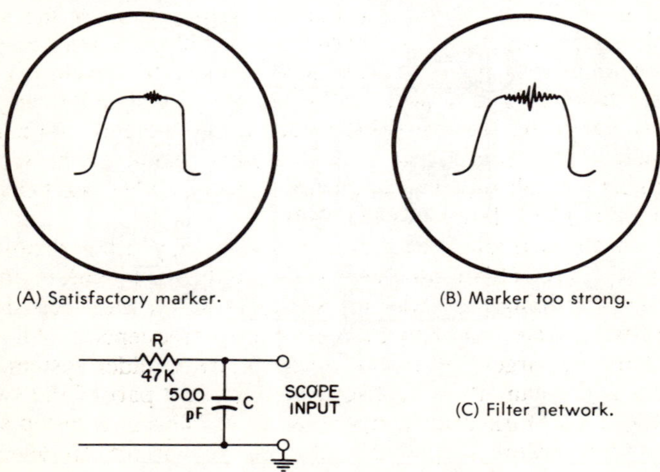

(A) Satisfactory marker.

(B) Marker too strong.

(C) Filter network.

Fig. 11-9. Marker amplitude variations.

increased to the overload point. Sometimes both of these measures do not reduce the relative size of the marker because of extremely sensitive amplifier circuits, signal leakage from the marker generator, or other reasons. The marker width can be decreased by the filter network shown in Fig. 11-9C if connected across the scope input terminals. With this network in place, the higher frequencies of the marker beat are bypassed; since the higher frequencies appear at the extremes of the marker, the marker width is reduced. Values of R and C are not critical, but the time constant RC should not be too large, or the response curve will be distorted.

The absorption marker system produces a dip, or indentation, in the response curve. Its action is similar to that of the traps usually found in video i-f strips of tv receivers; a high-Q resonant circuit absorbs energy at its resonant frequency from those circuits to which it is coupled. This loss of energy results in a dip in response at the marker frequency. Some technicians feel that this method causes less disturbance and less distortion of the response curve than the beat marker method. However, it does not seem so well suited to marking trap points on a response curve, since the amplifier gain at these points has already been greatly attenuated by the receiver traps.

Either marker system can be designed for tunable or fixed operation. For fixed operation, some more commonly used frequencies are chosen, and the instrument is designed to supply markers at those frequencies.

The marker system can either be built into the sweep generator unit or designed as a separate unit. If it is a built-in unit, the marker signal is applied to the receiver at the same point as the sweep signal; a separate marker unit poses the problem of marker injection methods. When an rf generator is used to develop a beat marker like the one shown in Fig. 11-9A, its signal strength is usually so strong that very little coupling is needed. Sometimes, just placing the output lead near a sensitive point in the receiver will inject enough signal to produce a marker. Other marker injection methods are shown in Fig. 11-10.

We have mentioned that beat markers may often distort the response curve. Other problems with this type of marker are the difficulty in seeing the exact marker location on the steep side of a response curve, and attenuation at trap frequencies. All these problems are largely overcome by the marker adder system. The marker adder can either be a separate unit or a part of the sweep-marker units. Marker adder operation is as follows: a sweep signal from the generator is applied to the receiver circuits to develop a response curve in the conventional manner; a portion of the sweep signal is applied separately to the marker adder unit, where it is

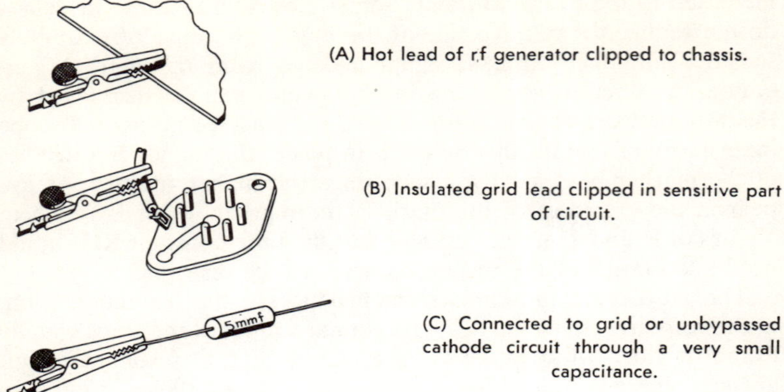

Fig. 11-10. An rf (or marker) generator signal may be injected by loosely coupling the signal to a receiver circuit.

beat with the marker signal and fed to a detector stage; the detected beat then receives any necessary amplification before being combined in the final stages of the adder with the response curve taken from the receiver and passed on to the oscilloscope for viewing.

Since the sweep signals for the receiver response curve and marker adder are both taken from the same sweep generator, the marker will appear at its proper location when combined with the response curve. The marker signal does not pass through the receiver circuits and, therefore, is not attenuated by the trap circuits or amplified at peak response frequencies. The marker remains at the same amplitude on the oscilloscope screen as it is tuned across the entire response curve.

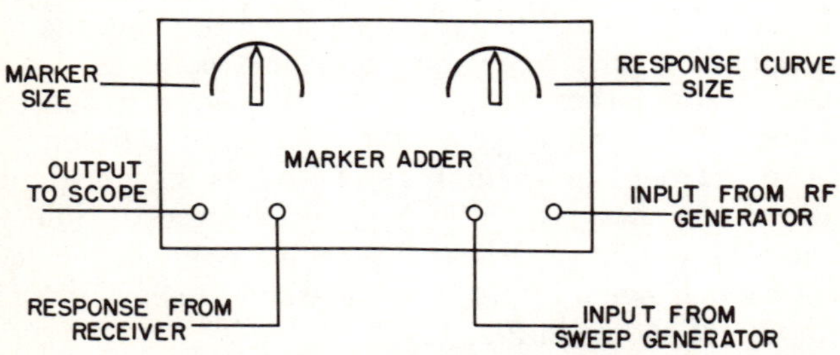

Fig. 11-11. Block diagram showing connections when using marker adder unit.

The block diagram of Fig. 11-11 shows the connections necessary to use a certain marker adder unit. The receiver response curve (Fig. 11-12) is taken from the same point as though the marker adder were not used.

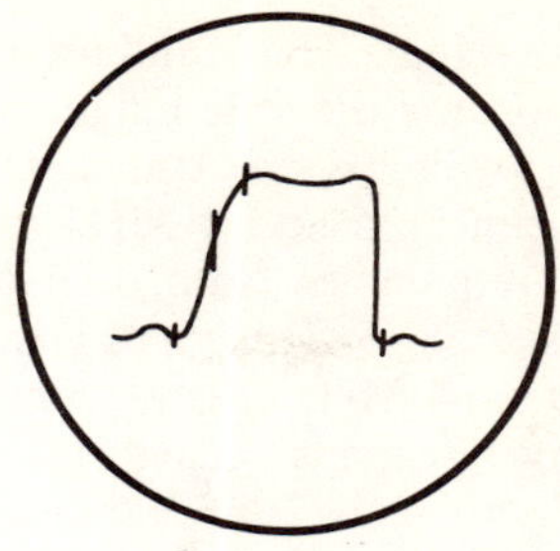

Fig. 11-12. Markers as they appear on response curve using a marker adder.

RESPONSE CURVE

Fig. 11-4 shows the equipment setup for obtaining a receiver response curve. The sweep generator supplies a signal that is continuously changing or sweeping through a range of frequencies. The extent of this range is governed by the sweep width control setting, and the center frequency of the sweep is governed by the main tuning control of the sweep generator. The output of a well-designed sweep generator should be flat over the entire sweep width, and, thus, a signal constantly changing in frequency but not in amplitude is applied to the receiver circuits. The scope response curve is a graph showing receiver amplification as vertical deflection and frequency variation as horizontal deflection.

Some oscilloscope operators are not entirely certain of the oscilloscope response needed to view a response curve of, for example, a video i-f strip. It seems to them that if a frequency of 25 MHz is represented in the response curve, the oscilloscope response should extend to 25 MHz. Although the response curve does represent 25 MHz, this frequency is absent in the signal applied to the oscilloscope. The response curve frequency is the same as the repetition rate of the generator sweep signal, usually 60 Hz. Thus, the signal applied to the scope during a video i-f response check resembles a 60-hertz, square-wave signal and can be viewed with a scope having a response of a few cycles per second to a few thousand cycles per second. If a sweep signal is viewed without a detector, the scope response should extend to the sweep signal frequencies.

An idealized video i-f response curve is shown in Fig. 11-13. This response curve will give a flat output from the standard tv

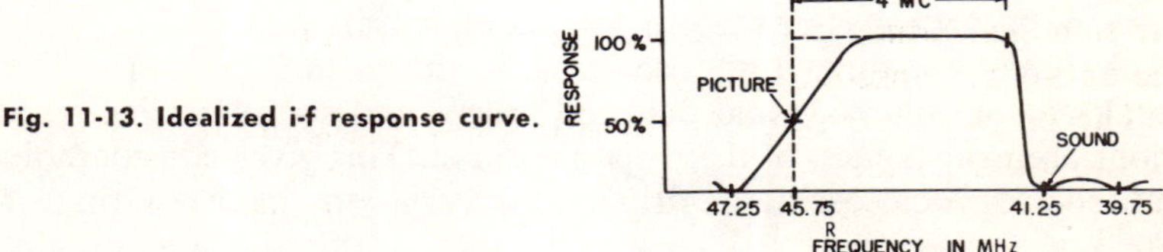

Fig. 11-13. Idealized i-f response curve.

video signal. The standard tv signal uses vestigial sideband transmission for the video signal. In other words, part of one sideband (the lower) is not transmitted at full strength. Video frequencies between 1.25 and 4 MHz are more or less completely attenuated, and frequencies from 0.75 to 1.25 MHz are partly attenuated in this sideband. This gives a transmitted bandwidth of approximately 5 MHz (4-MHz upper sideband plus 1-MHz lower sideband). If the receiver response curve is a rectangle covering these same 5 MHz, the frequencies from 0 to 1 MHz make a double contribution to the overall response (one for the upper and one for the lower sideband), and these frequencies are overemphasized. The response curve of Fig. 11-13 gives the same amplification as a rectangular response curve 4 MHz wide and is much easier to attain. The total contribution of the upper and lower sidebands at frequencies from 0 to 1 MHz equals that of the upper sideband above these frequencies; the total overall response is practically flat from 0 to 4 MHz.

The response is kept low at 39.75 MHz, 41.25 MHz and 47.25 MHz to prevent these frequencies from reaching the picture tube, where they would interfere with the desired picture. These intermediate frequencies correspond to adjacent picture, associated sound, and adjacent sound frequencies and are the most likely points for interference to develop. The response at these points is attenuated by traps in the receiver circuit. The sound i-f trap at 41.25 MHz not only keeps sound frequencies out of the picture, but also helps shape the response curve at the upper video frequencies. The response at undesired carrier frequencies generally should be 30 dB down from peak response. This response corresponds to 3 per cent of maximum gain. Since adjacent channels are not assigned to the same locality, adjacent channel traps are not always included in receiver design, but may be beneficial for fringe area reception.

The idealized response curve of Fig. 11-13 is seldom encountered in commercial tv receivers. Usually, the peak response will not be as wide or as flat. The alignment technician may get more gain from video i-f amplifiers by aligning for more gain and less bandwidth. Also, running the video carrier higher up the slope of the response curve will give more gain to the lower video frequencies. Some of the higher video frequencies will be lost, of course, which means less fine detail in the picture, but the additional video gain may be just what is needed in a weak signal area.

An idealized tuner response curve is shown in Fig. 11-14. The peak response is wide and flat, and video- and sound-carrier positions are near the top of the response curve. This gives considerable latitude for local oscillator drift or for variations in the setting of the fine tuning control. The actual tuner response curve usually

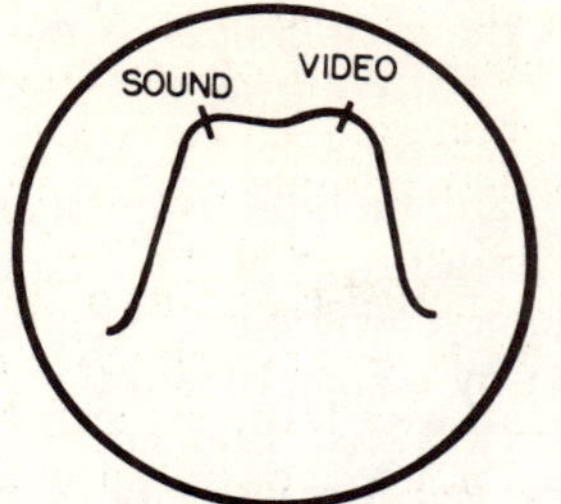

Fig. 11-14. Idealized tuner response curve.

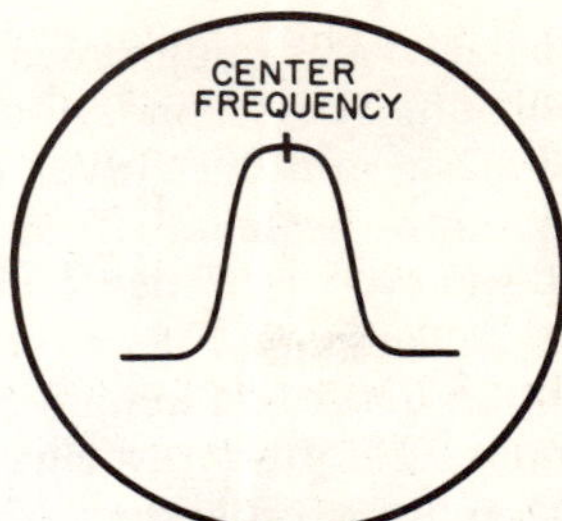

Fig. 11-15. An fm sound i-f response curve.

differs from the ideal, which is logical when you consider that the tuner is designed to receive many channels. Usually, the response will dip between the two carrier positions, and the carriers may be below peak response. At different channel settings the top of the response may slope at different degrees. The manufacturer's alignment instructions will usually state the degree of slant, amount of dip, and lowest permissible position on the response curve for the video and sound carriers.

Actual and theoretical response curves for fm sound i-f amplifiers correspond quite closely. Fig. 11-15 shows the response desired when adjusting the sound take-off and sound i-f transformers. Fig. 11-16 shows what response curve to obtain at the detector output. This curve applies to both ratio-detector and discriminator circuits. The marker indicates the center of the i-f band, 4.5 MHz for tv receivers and 10.7 MHz for most fm broadcast receivers. Recommended bandwidth of the curve of Fig. 11-15 and the straight-line portion of Fig. 11-16 is 100 kHz for tv receivers and 200 kHz for fm receivers. This bandwidth is necessary for distortionless reception at maximum modulation.

During alignment of the detector transformer, the primary adjustment controls the amplitude of the response curve (Fig. 11-16),

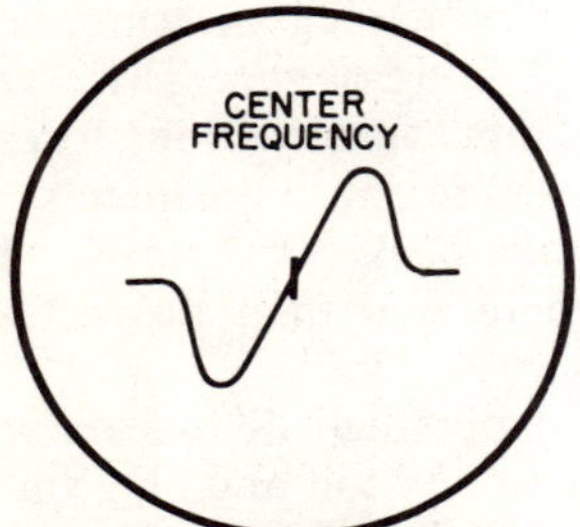

Fig. 11-16. An fm-discriminator or ratio-detector response curve.

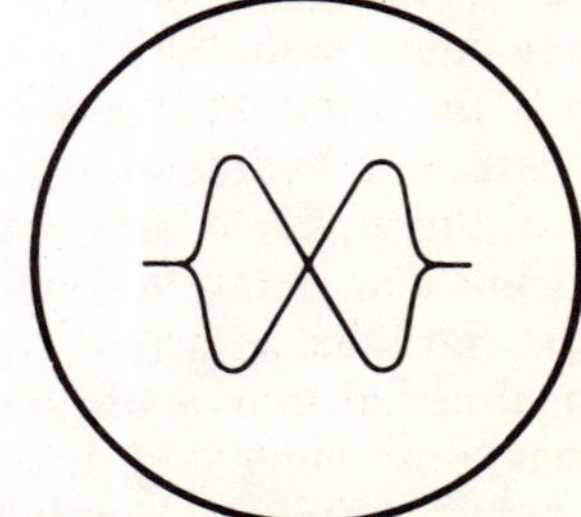

Fig. 11-17. Double response curve of fm detector.

and the secondary adjustment controls the straightness of the middle portion. The secondary adjustment also controls the centering of the marker on this curve. These two adjustments are interactive; that is, one adjustment affects the other, and their adjustment should be repeated several times, ending with adjustment of the secondary.

The S curve of Fig. 11-16 is obtained when the oscilloscope horizontal deflection amplifier is driven by a signal from the sweep generator. The butterfly curve of Fig. 11-17 is obtained by using the internal sweep system of the scope, set for a sawtooth sweep twice the frequency of the generator sweep. This type of curve is preferred by some as an aid in judging marker centering.

REVERSED AND INVERTED CURVES

The alignment technician may be disturbed at times by a response curve that does not agree with the curve pictured in the alignment literature. This may be caused by reversal or inversion of nonsymmetrical curves. Thus, whereas the video marker may appear on the left-hand slope of the response curve in the literature, it appears on the right-hand slope on the scope. This is no cause for concern; it merely indicates that the final horizontal sweep voltage applied to the oscilloscope deflection plates is of opposite polarity to the one used for the alignment example. It can be the result of a different number of stages in the two oscilloscopes or a difference in the two sweep generators. As long as the different points on the response curve can be identified, the direction of sweep should not affect the final results of the alignment.

SOME ASPECTS OF ALIGNMENT ADJUSTMENTS

When a number of adjustments are to be made, say, during alignment of a complete video i-f strip, the technician may find that as the alignment progresses, the receiver breaks into oscillations, preventing further adjustment of that particular slug or trimmer. A common cause for this behavior is the peaking of successive stages to nearly the same frequency. Some alignment instructions forestall this possibility by recommending prealignment adjustments; for example, "turn slug A all the way in, then back out 3 turns." This adjustment will place the slugs somewhere near their proper positions so that little adjustment is necessary.

Sometimes a desired response can be obtained at two different positions of a tuning slug. If there are another coil and slug on the same form, the recommended position will place the two slugs farthest apart. This position results in minimum reaction between the two alignment adjustments.

A tuning wand (Fig. 11-18) is a very useful tool for alignment purposes. Two slugs are mounted on opposite ends of a rod of a material which will not affect circuit operation when placed next to circuit components. One slug is brass and the other is of iron

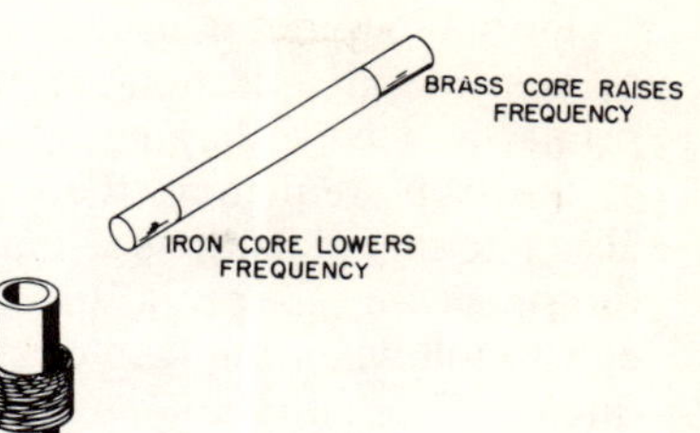

Fig. 11-18. Tuning wand placed next to coil simulates alignment adjustment.

composition, similar to the tuning slugs. When the iron end of the wand is brought next to a coil, the inductance of the coil is increased, lowering the resonant frequency. Conversely, when the brass slug is placed next to the coil, the inductance of the coil decreases, raising the resonant frequency. The response curve is ob-

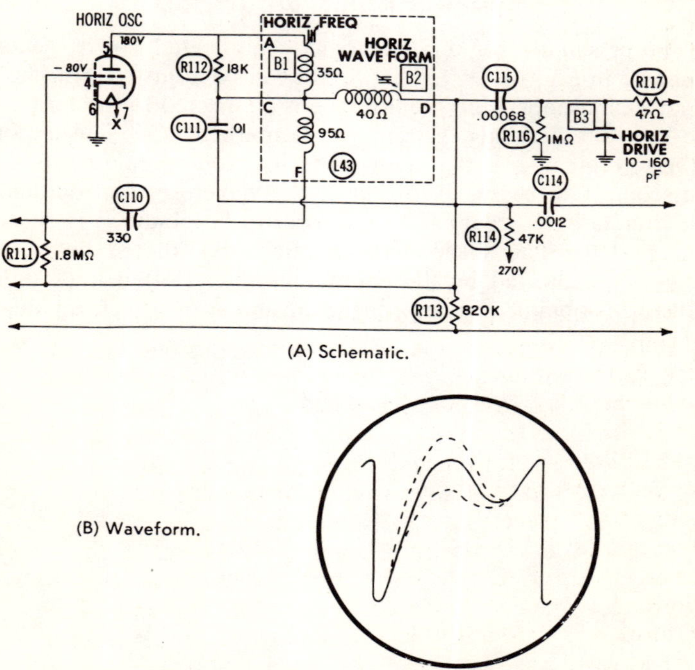

(A) Schematic.

(B) Waveform.

Fig. 11-19. Synchroguide horizontal oscillator circuit and waveforms.

served, meanwhile, so that the effect of an alignment adjustment can be determined in this manner before the adjustment is made.

TRAP ALIGNMENT

Trap alignment is usually made to reduce the response as much as possible at a certain frequency, although sometimes it may be primarily a waveshaping operation. As the response is attenuated at the trap frequency, the technician will find that the marker is also attenuated (unless a marker adder system is used) and may disappear unless special precautions are taken. This makes it difficult to tell when the trap circuit is tuned exactly to the marker frequency. The situation can be improved in several ways: the sweep signal strength can be reduced and the scope vertical gain increased, or the marker strength can be increased; the sweep width can be decreased by the sweep generator control. This latter step will expand the trap region of the response curve on the scope. The curve can also be expanded by the horizontal gain control of the scope. All these measures help make the trap marker more visible.

SYNCHROGUIDE WAVEFORM

The schematic for a Synchroguide horizontal sweep generator is shown in Fig. 11-19A. This circuit can be adjusted with the help of an oscilloscope. The oscilloscope is connected to the terminal of the horizontal frequency transformer marked "C." The important point to remember is that a low-capacity probe should be used with the scope. Otherwise, distortion of the waveform and detuning of the circuits may occur. The waveform to be obtained is shown in Fig. 11-19B. The sharp and smooth peaks should be of equal height, as indicated by the continuous line. Dotted lines indicate waveforms obtained by improper adjustment of waveform slug B2.

Chapter 12

Signal Tracing and Other Applications

Signal tracing is one of several methods that may be used to service electronic equipment. Certain defects may be just as easily discovered by voltage and resistance measurements and by tube substitution, whereas others will be found with more ease and certainty by signal-tracing methods. The latter defects are more likely to be of the halfway type—that is, the circuit functions, but not perfectly. Some circuits in a receiver may check within the normal tolerance limits of specified resistances and voltages and yet not perform their intended duty. (For example, an oscillator circuit may be off frequency or have a poor waveshape.) Other circuits are passive except for their effect on a signal, as for example, sync clippers and separators in a television receiver. The oscilloscope makes a fine instrument for locating defects in such circuits because it can show amplitude and waveshape at the same time.

To use the oscilloscope for signal tracing, a signal of proper characteristics is applied to the circuits under question, and the oscilloscope gives a point-by-point indication of any change in the appearance of the signal. In some instances, the circuit itself furnishes a signal that can be traced. If the signal is absent where it should be present, present where it should be absent, or abnormal in amplitude or waveshape, it indicates that some defect exists between signal generator and oscilloscope. By decreasing the gap between generator and scope, the defective stage can be located. Further checks with a voltmeter or ohmmeter will usually locate the defective component.

For signal-tracing purposes, it is convenient to divide a radio, either a-m or fm, into about four sections. They are (1) the rf and converter section, (2) the i-f amplifier, (3) the sound detector or second detector, and (4) the audio amplifier. It will be noticed that this division groups the circuits according to the type of signal passing through them. For example, consider a standard broadcast a-m receiver tuned to receive a 1000-kHz carrier with audio modulation. The signal is unchanged (except for possible changes in amplitude) until it reaches the converter stage. There, it is converted to the intermediate frequency, usually 455 kHz, with audio modulation still present. The 455-kHz signal is amplified by the i-f stages and then applied to the detector stage. From that stage on, it is an audio signal.

We see, then, that at least three different types of signal can be used for signal-tracing the a-m radio. The one chosen will depend on the point of application. To trace through the entire audio section, an audio signal is applied at the output of the sound detector; to trace through the i-f and audio stages, an audio-modulated i-f signal is applied at the output of the converter stage; and to trace through the entire receiver, an audio-modulated rf signal is applied to the antenna terminals. These signal application points are indicated on the block diagram of Fig. 12-1 as points 1, 2, and 3,

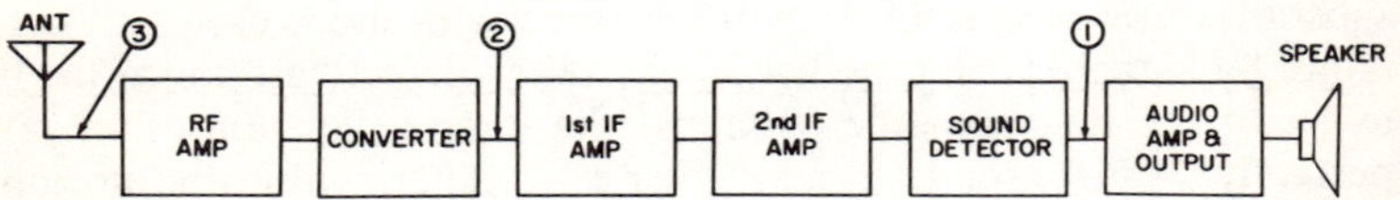

Fig. 12-1. Block diagram of a radio receiver showing points where different types of signal are injected for signal tracing.

respectively. It is possible to apply an audio-modulated i-f signal at the input stage of many receivers and force a signal through the receiver, but this requires a stronger signal because the rf stages are not designed to amplify these frequencies. Receivers with very selective rf stages would be difficult to trace in this manner.

Points 1, 2, and 3 of Fig. 12-1 are not the only points where a signal can be injected for tracing purposes. They are merely the points where the type of signal should be changed. A signal can be injected anywhere that it is convenient or desirable to do so. Likewise, the scope can be connected to any point for viewing, depending upon the number of stages to be included between signal generator and scope. With standard broadcast a-m receivers, it is not necessary to insert a detector between scope input and receiver since most present-day scopes have a vertical-amplifier response covering all the receiver frequencies. Fig. 12-2 shows the two types of patterns to expect under normal conditions with a modulated rf signal applied. The nature of the signal ahead of the sound detector is shown in Fig. 12-2A, and the signal following the sound detector is shown at Fig. 12-2B. The first example is similar to the input signal—that is, it is one sine-wave signal modulated by another. The second example shows the modulating signal alone—the carrier signal has been removed by the demodulation circuits. As was stated, these are normal patterns. Under abnormal conditions, they will probably be quite different. In circumstances where there is very little gain between generator and oscilloscope, the maximum oscilloscope gain may be necessary to obtain a usable pattern. This will bring up the hum and noise level in the pattern.

The block diagram of Fig. 12-1 will illustrate an fm receiver as well as an a-m receiver. Frequencies involved will be different; the sound detector operates by a different principle, but otherwise, the function of each block is quite similar in either receiver. It is not necessary to use an fm signal to signal-trace the receiver. An

(A) Modulated rf or if.

(B) Demodulated signal.

Fig. 12-2. Response patterns obtained in signal tracing an a-m radio receiver.

amplitude-modulated rf signal will be converted, amplified, and will even pass the fm detector section. It is best to detune the rf generator slightly off the center frequency since discriminator circuits are not sensitive to amplitude changes at this frequency. The rf and i-f signals used to signal-trace an fm receiver will be above the frequency range of most oscilloscopes; so, a demodulator probe must be used when the scope is connected ahead of the fm detector. The applied signals should be kept low in amplitude to avoid driving any stage to limiting action.

If the technician wishes, he can use an fm signal generator instead of the a-m generator. This will result in response curves similar to those obtained during an alignment procedure.

A television receiver can be divided into sections in much the same manner as the a-m or fm receiver: rf amplifier, converter, video and sound i-f amplifiers, and video and audio amplifiers. In addition, there are the circuits pertaining to vertical- and horizontal-sweep generation and synchronization. Fig. 12-3 shows these various sections in a block diagram. The number of video i-f and sound i-f amplifier stages will vary with different makes and models of receivers and, therefore, will not necessarily agree with the number shown here.

One of the simplest ways of tracing the rf and video i-f stages is to use a television broadcast station as a signal source, if one can be received strongly. A demodulator probe should be used with the

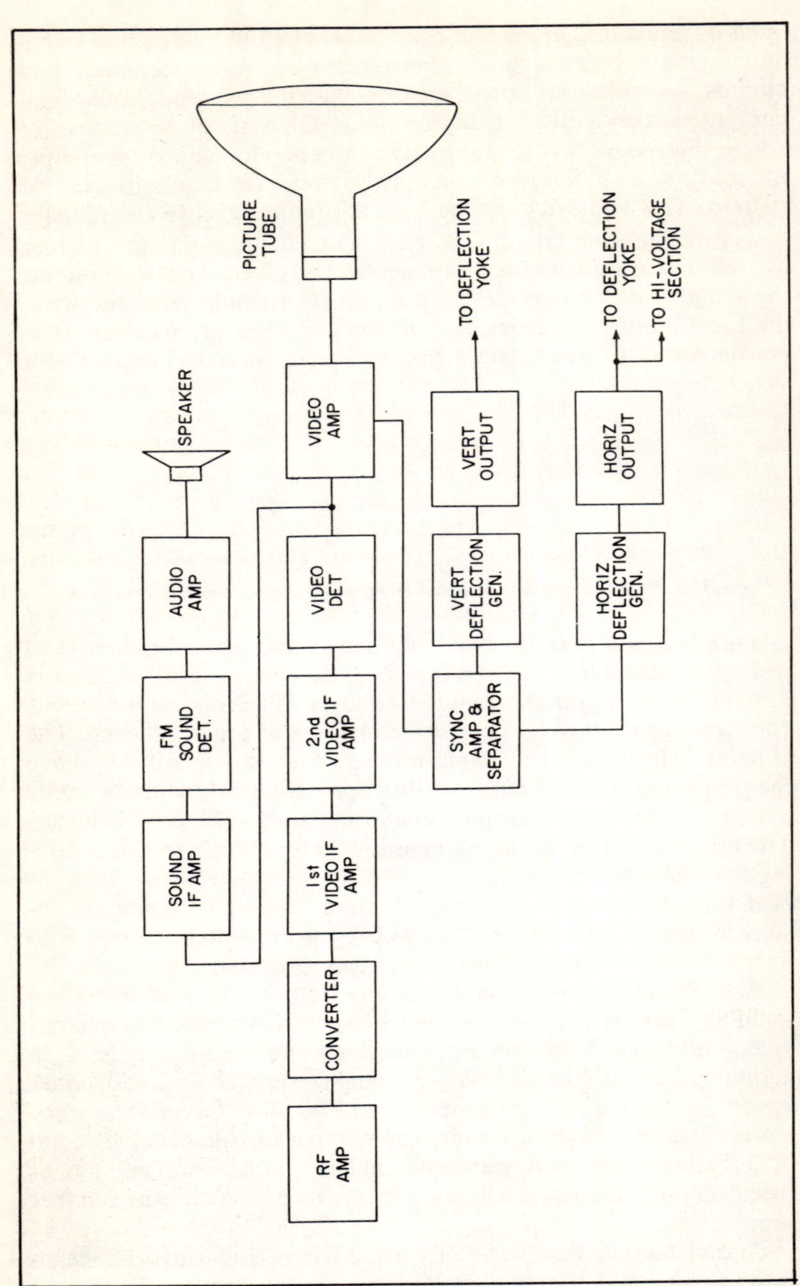

Fig. 12-3. Block diagram of typical tv receiver.

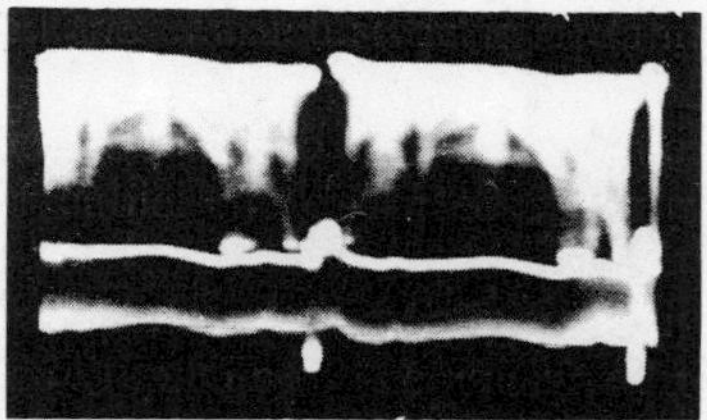

Fig. 12-4. Normal video signal at video detector.

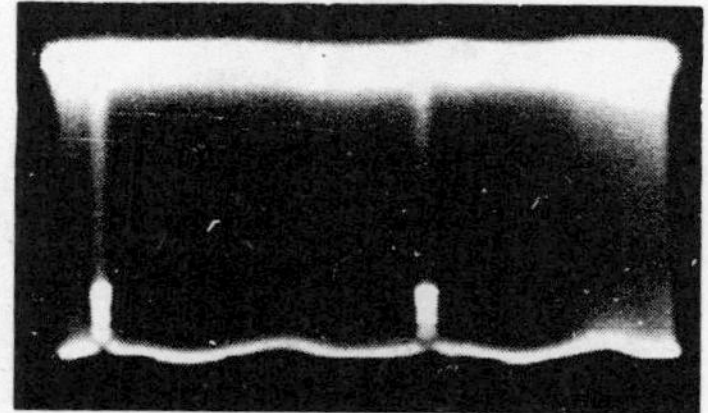

Fig. 12-5. Video signal after one stage of sync separation. Most of the video information has been removed.

oscilloscope. The detected video signal normally appears as in Fig. 12-4 and, of course, should get smaller in amplitude as one proceeds toward the rf section of the receiver. The oscilloscope sweep can be set to synchronize with either the vertical or horizontal sync pulses, but it is usually easier to use the vertical sync rate. The sweep rate for Fig. 12-4 was 30 Hz. The waveform of Fig. 12-4 should also be visible through the video amplifier, up to the picture tube. It is unnecessary to use the demodulator probe in the stages following the video detector.

A portion of the video signal is applied to the sync amplifier and separator stages, where the video information is removed from the signal leaving the vertical and horizontal sync signals. Fig. 12-5 shows the video signal after one stage of sync separation. Most of the video information has been removed, and will be removed completely in the next stage. The vertical sync signal is passed to the integrator network, and the horizontal pulses are passed to the horizontal-oscillator section. The output of the integrator network appears at Fig. 12-6. The vertical blocking oscillator was temporarily disabled to obtain this picture. The synchronized waveform at the vertical-output stage is shown in Fig. 12-7.

Fig. 12-8 shows the waveforms obtained at the horizontal phase detector when the receiver is synchronized. The signal fed back from the horizontal-output stage is shown in Fig. 12-8A, and the outputs from the two halves of the phase detector are shown in

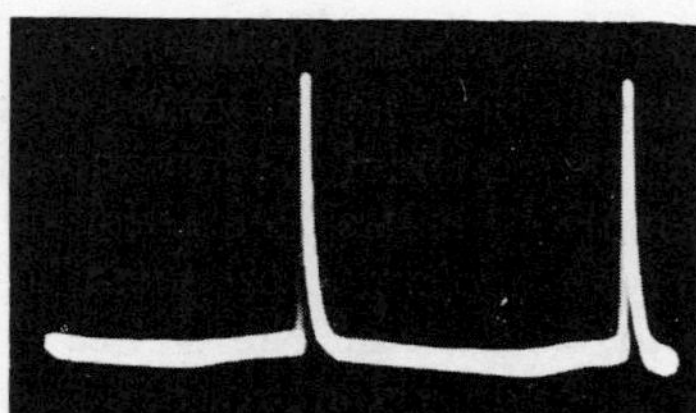

Fig. 12-6. Vertical pulses from the output of the vertical integrator network.

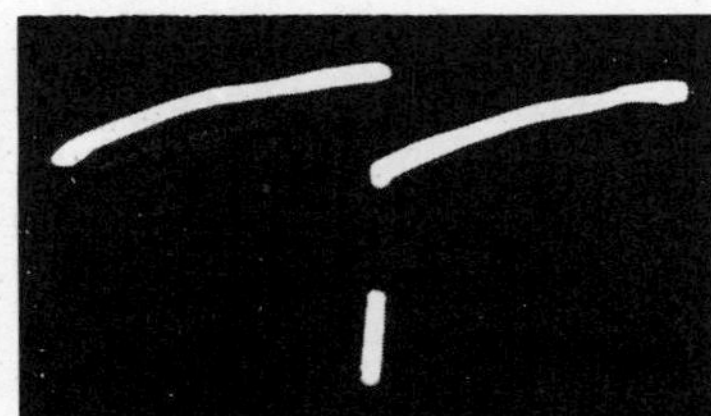

Fig. 12-7. Vertical deflection waveform from vertical output stage.

(A) Input signal to phase detector.

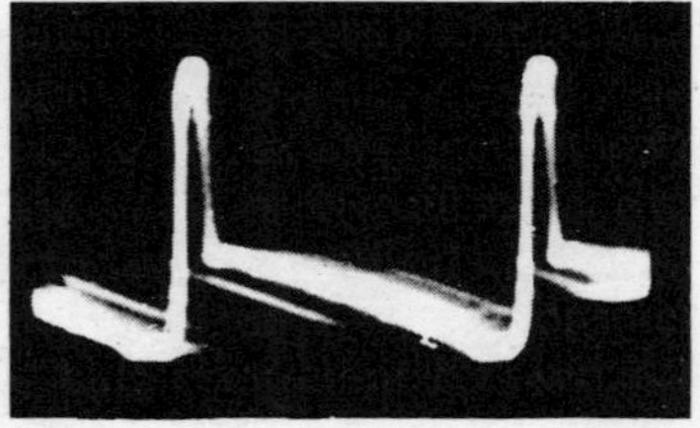

(B) Output from one half of detector.

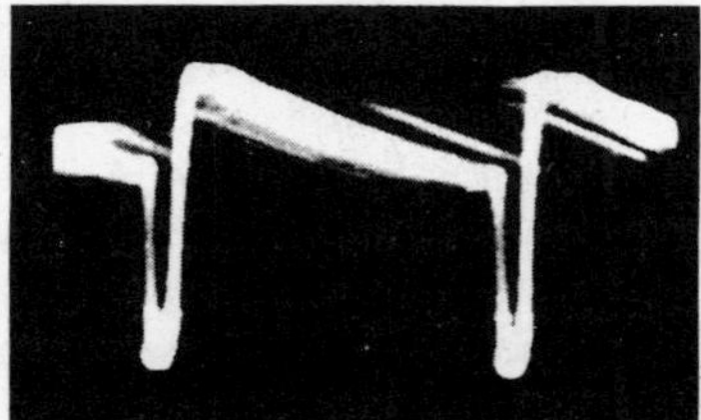

(C) Output from other half of detector.

Fig. 12-8. Waveforms obtained at the horizontal phase detector.

Figs. 12-8B and 12-8C. One of the waveforms from the horizontal multivibrator stage is shown in Fig. 12-9. This waveform was taken at the second grid of the multivibrator tube.

Signal-tracing procedure for the sound system of a tv receiver is practically the same as for an fm broadcast receiver. The principal difference between the two systems is in the sound intermediate frequency—4.5 MHz for television and 10.7 MHz for fm broadcasts.

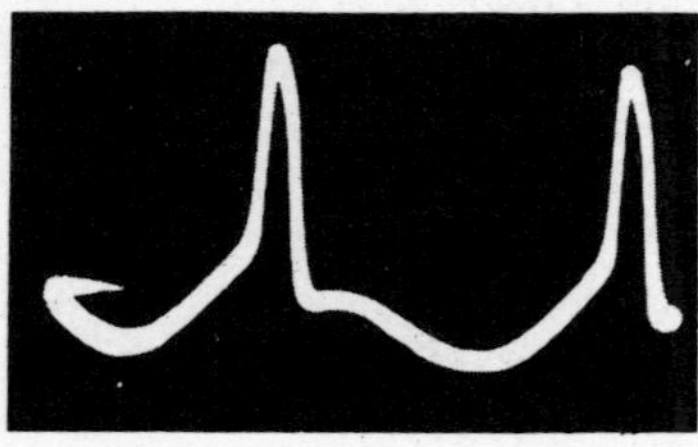

Fig. 12-9. Waveform obtained at the second grid of the horizontal multivibrator stage.

CHECKING DISTORTION OF AMPLIFIERS

The oscilloscope can be used for certain types of distortion checks in amplifiers. (We refer principally to audio amplifiers.) Distortion in an amplifier can be described as a condition that causes the output waveform to be different from the input waveform, neglecting differences in overall size. If the difference is great enough, it can be observed on the oscilloscope. Several types of distortion will affect amplifiers: amplitude, harmonic, intermodulation, transient, phase, and frequency distortion.

Amplitude distortion is caused by some nonlinear condition in the amplifier and results in a difference in gain for signals of different amplitudes. Harmonic distortion and intermodulation distortion are a direct result of amplitude distortion. Harmonic distortion causes harmonics to appear in the output when a pure sine-wave input voltage is amplified. Intermodulation distortion causes two input signals to interact so that one modulates the other.

Transient distortion results in improper reproduction of sudden changes in input signal. Phase distortion alters the phase angle between a fundamental and some harmonic or between any two frequencies in a complex waveform. Frequency distortion is the unequal amplification of different frequencies. If we refer to a response curve of an amplifier, frequency distortion will occur at those frequencies indicated by dips or peaks in the curve.

It would be difficult to determine the types or percentage of distortion in an amplifier by inspection of the output waveform. Small percentages of certain types of distortion will not change the appearance of a sine wave too much, but if the peaks are rounded or clipped by overload conditions in an amplifier, the distortion is evident. The value of a scope distortion check is that changes in distortion are readily seen. Thus, the scope probe can be moved from point to point in an effort to locate the source of distortion. If the distortion is localized in this manner to a certain stage, the circuit components in that stage can be checked for such defects as opens, shorts, or changes in value. As previously pointed out, an electronic switch can be used advantageously when two different stages in an amplifier are compared.

Multiple feedback paths in an amplifier make it more difficult to perform oscilloscope checks because the feedback action may be upset by the added capacitance of the scope input circuit.

The oscilloscope can be used as an aid to amplifier checking with an intermodulation meter. Connected across the output load of the amplifier, the scope will show when the clipping or overload point is reached. At the same time, the scope can be used as a power output indicator if reference is made to a chart listing peak-to-peak voltages for watts output. The chart must have a column for the rated output impedance of the amplifier (usually 8 or 16 ohms).

BALANCE OF PUSH-PULL AMPLIFIER STAGES

Some amplifiers provide for balancing the push-pull stages. Ac balance can be checked with the oscilloscope. If the push-pull output stage has a common cathode resistor as shown in Fig. 12-10, this is a convenient point to connect the scope. A signal of normal amplitude is applied to the amplifier, and the balance adjustment is made for minimum indication on the scope. This adjustment can be quite sensitive if the oscilloscope gain is set at maximum. If the adjustment does not reach a null or minimum point, it can be helped sometimes by trading the output tubes in their sockets.

Another method of checking ac balance is to connect the oscilloscope, first to one output plate and then to the other, and compare the signal amplitude at each plate. Here again, the electronic switch can be used advantageously because the signals are compared simultaneously instead of alternately.

Certain unconventional output circuits may not be suited to either of the balance check methods just mentioned. The technician should follow the recommendations of the manufacturer, if any are offered, for ac balancing of these circuits.

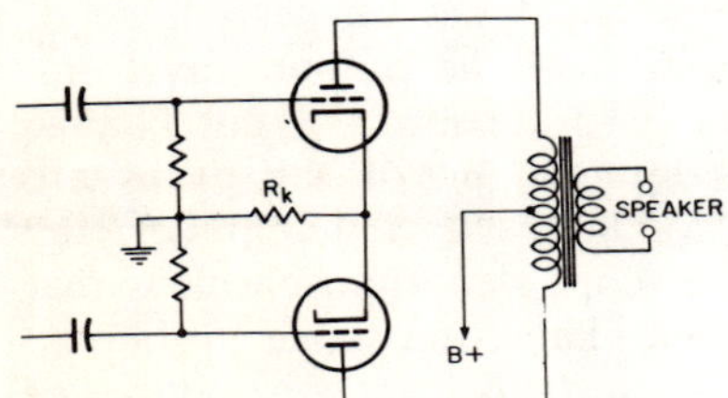

Fig. 12-10. Oscilloscope can be connected across common cathode resistor R_k for ac balance adjustment.

TUBE CHARACTERISTIC CURVES

Fig. 12-11 shows the connection methods that will enable the oscilloscope operator to view one type of electron tube characteristic curve. The curve developed is an operating characteristic of the tube under the load conditions shown by the circuit. A number of response curves obtained from a setup like that of Fig. 12-11 are shown in Figs. 12-12 through 12-15. The electron tube used was a medium-mu triode. The plate voltage supply E_b was 285 volts. R_g was 470K ohms, R_L was 100K ohms, and R_k was 3300 ohms. Vertical- and horizontal-gain controls were adjusted to main-

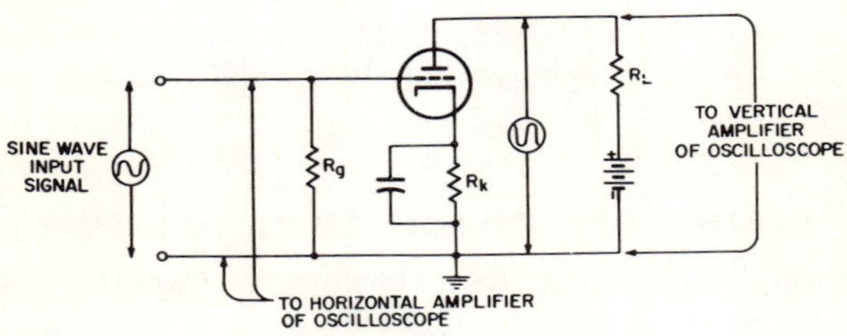

Fig. 12-11. Method of connecting oscilloscope to an amplifier stage to view the operating characteristic curve.

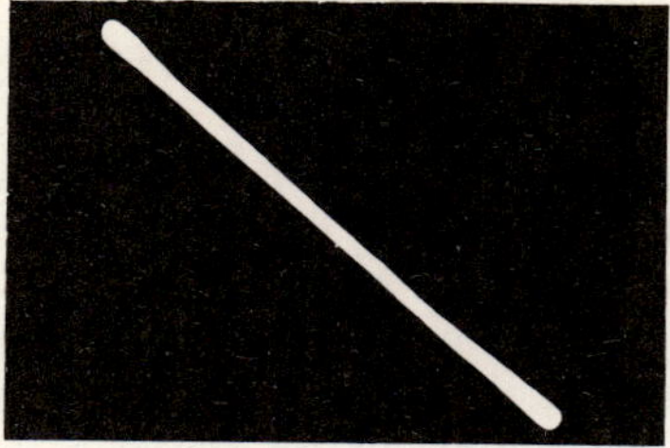

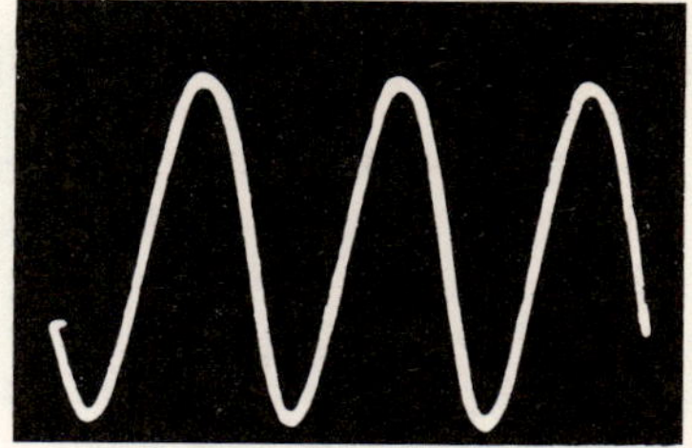

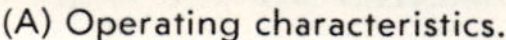

(A) Operating characteristics. (B) Sine-wave output.

Fig. 12-12. Triode response with small signal input.

tain the response curve at a convenient size. Notice that the characteristic curves slant upward toward the left, rather than toward the right as do many curves found in tube manuals. The following explanation will show why this is true: (1) For a positive-going signal, vertical deflection of the oscilloscope beam was

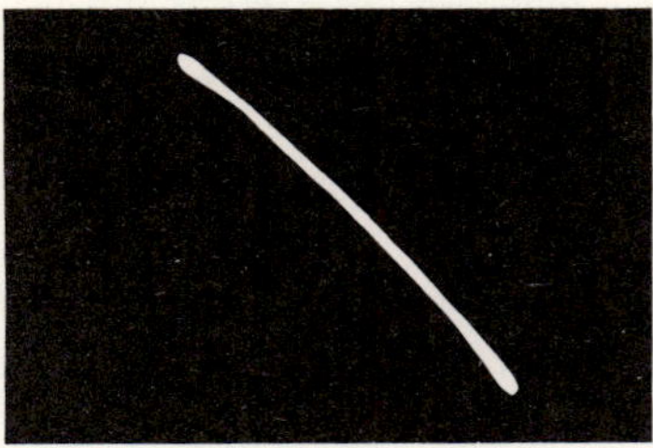

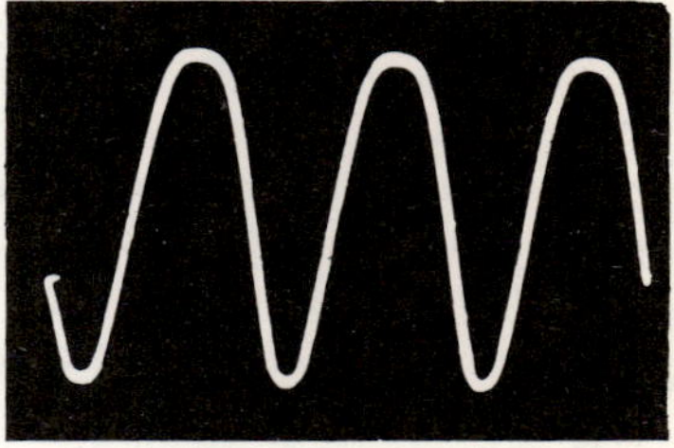

(A) Operating characteristics. (B) Sine-wave output.

Fig. 12-13. Triode response with small overload.

upward, and horizontal deflection was to the right. (2) The vertical input of the oscilloscope was connected to the plate, and vertical deflection, therefore, corresponds to instantaneous plate voltage. Because of voltage drop across plate load resistor R_L, the plate voltage decreases (becomes more negative) as the grid signal increases. Curves shown in tube manuals usually show plate current

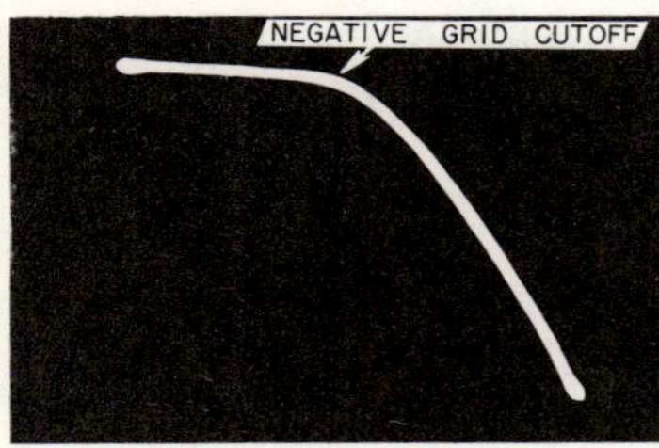

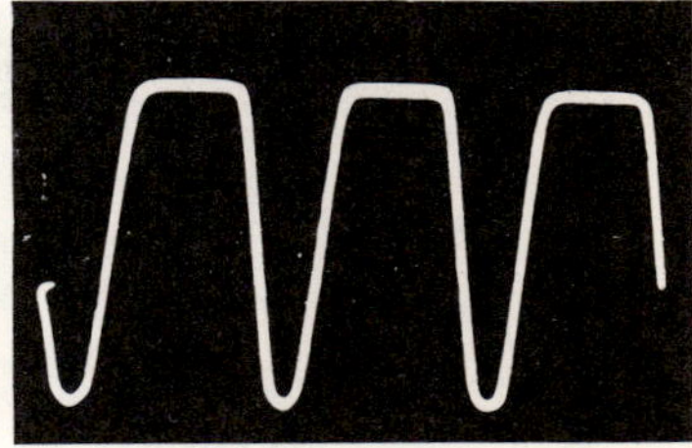

(A) Operating characteristics. (B) Sine-wave output.

Fig. 12-14. Triode response with large overload.

rather than plate voltage, and plate current increases as grid voltage increases.

The straight line response of Fig. 12-12A was obtained with a small signal input. The straight line, of course, indicates linear operation. Fig. 12-12B bears this out by showing how the sine-wave input signal is reproduced undistorted at the plate of the amplifier stage. To obtain this response curve, the oscilloscope controls were merely changed from external to internal sweep operation; input signal and connection points were not changed.

When the input signal strength was greatly increased, the operating characteristic of Fig. 12-13A was obtained. The response curve was also increased in size by a proportionate amount, and it was necessary to readjust the vertical and horizontal controls to return the curve to its original size. As shown by this curve, linear operating conditions have been exceeded, and the distorted output waveform of Fig. 12-13B results.

A still greater increase in signal strength produced the greatly distorted waveforms of Fig. 12-14. The arrow in Fig. 12-14A indicates the point of negative grid cutoff. The plate current has been reduced to zero by a highly negative excursion of the grid signal and the plate voltage reaches its highest value under these conditions. Making the grid still more negative has no further effect on the plate voltage, and therefore, the curve remains flat for the remainder of the cycle.

Response curves similar to those just mentioned can be obtained by holding the input signal constant and varying other circuit parameters, like grid bias or plate voltage. Fig. 12-15 is a diagram of the results obtained by applying separate values of grid bias to the circuit of Fig. 12-11. The grid return was opened at the grounded end of R_g, and a bias supply was inserted between R_g and ground. Curves 1, 2, and 3 correspond to negative bias voltages of 3, 9, and 13 volts, respectively. The degree of slant for each curve indicates the relative amplification obtained for each bias value. Curve 1 indicates the highest amplification factor. If the gain of both horizontal and vertical amplifiers of the oscilloscope were known, the absolute value of the amplification factor could be calculated.

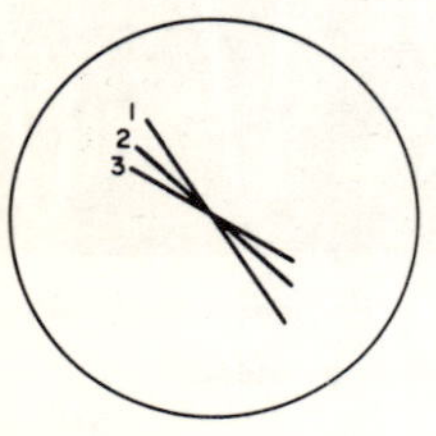

Fig. 12-15. Changes in operating characteristic obtained by varying the grid bias of an amplifier stage.

These examples show only a few of the possibilities for analysis of tube characteristics with an oscilloscope. Such analysis might be used as a practical approach to the design of amplifier stages, where it is desirable to select the best circuit values by the "cut and try" procedure. The exact effect of each selection can be seen and evaluated.

MEASUREMENT OF SMALL AC SIGNALS

Oscilloscopes are being designed with higher sensitivities and wider frequency-response characteristics than earlier models. These properties, coupled with the fact that the signal waveshape is shown and can be measured even though of complex form, make the oscilloscope a superior instrument for measuring small ac voltages. For example, a manufacturer states that one model has a sensitivity of 5 millivolts for full-scale deflection. This permits direct measurements from low-output transducers without the use of a preamplifier.

CURRENT MEASUREMENT

Beam deflection in an oscilloscope is a direct result of voltages applied to the deflection plates. The oscilloscope is, therefore, a voltage-indicating device. However, it can be used to measure current under certain conditions. The current must first be made to provide a voltage indication, and this can be done by passing the current through a resistance. The value of R must be known, and E can be measured with the oscilloscope; then I can be obtained from Ohm's law: $I = E/R$.

A few precautions should be observed: the value of R should be kept small, compared to the impedance of the circuit to which it is connected. If R is too large, it will tend to reduce the total current, and the scope indication will not be the true value for the original circuit. The resistor should be as free from reactance as possible—that is, it should have a minimum of inductance or capacitance to avoid phase shift complications.

Index